T0312053

Automotive Process Audits

Practical Quality of the Future: What It Takes to Be Best in Class (BIC)

Series Editor:
D. H. Stamatis
President of Contemporary Consultants, MI, USA

Quality Assurance
Applying Methodologies for Launching New Products, Services,
and Customer Satisfaction
D. H. Stamatis

Advanced Product Quality Planning
The Road to Success
D. H. Stamatis

Automotive Audits
Principles and Practices
D. H. Stamatis

Automotive Process Audits
Preparations and Tools
D. H. Stamatis

For more information about this series, please visit: https://www.routledge.com/ Practical-Quality-of-the-Future/book-series/PRAQUALFUT

Automotive Process Audits
Preparations and Tools

D. H. Stamatis

CRC Press
Taylor & Francis Group
Boca Raton London New York

CRC Press is an imprint of the
Taylor & Francis Group, an **informa** business

First edition published 2021
by CRC Press
6000 Broken Sound Parkway NW, Suite 300, Boca Raton, FL 33487-2742

and by CRC Press
2 Park Square, Milton Park, Abingdon, Oxon, OX14 4RN

Library of Congress Cataloging-in-Publication Data
Names: Stamatis, D. H., 1947- author.
Title: Automotive process audits : preparations and tools / D.H. Stamatis.
Description: First edition. | Boca Raton : CRC Press, 2021. |
Series: Practical quality of the future: what it takes to be best in class (BIC) |
Includes bibliographical references and index.
Identifiers: LCCN 2020051486 (print) | LCCN 2020051487 (ebook) |
ISBN 9780367759391 (hbk) | ISBN 9781003164715 (ebk)
Subjects: LCSH: Automobiles—Design and construction—Quality control. |
Manufacturing processes—Quality control. | Automobile industry and trade—Auditing.
Classification: LCC TL278.5 .S8293 2021 (print) | LCC TL278.5 (ebook) |
DDC 629.222068/5—dc23
LC record available at https://lccn.loc.gov/2020051486
LC ebook record available at https://lccn.loc.gov/2020051487

ISBN: 978-0-367-75939-1 (hbk)
ISBN: 978-1-003-16471-5 (ebk)

Typeset in Times
by codeMantra

To all those who dedicate themselves to improve the quality culture *of an organization through process auditing*

Contents

List of Figures

List of Tables

Preface

The field of quality as time progresses demands "more" quality in all areas of design, manufacturing, and service. This means nonconformances must be removed from the delivered part or service from the supplier base to the customer. The aim is to ever-satisfy the customer.

Over the years, the determination of "quality" has taken many roads, such as visual inspection, MIL-STDS, industrial standards, individual organizational standards, and since the late 1980s international standards. It seems that the international standards are being accepted by many worldwide organizations and are known as the ISO standards published by the International Standards Organization, based in Zurich, Switzerland.

The international standards known as the ISO 9001, ISO 14001, ISO 18001, or ISO 45001 series (and many others for different areas) offer a basic system of quality, environment, or occupational, health, and safety opportunities to build that quality, environment, and safety throughout "a" given organization. In fact, the standards have become the blocks of quality by which "a given" organization can achieve total quality management (TQM) and world-class quality. The standards offer a common foundation for all to work from and at the same time allow for individuality. In fact, these standards have accelerated the use of old methodologies and new methodologies in identifying, controlling, and in some cases eliminating nonconformances. Some of these are the classical seven tools of quality, statistical process control (SPC), Six Sigma, advanced statistical analyses, and auditing.

However, because the standards are for all intents and purposes generic in nature, many industries and organizations have developed their own. Typical standards that fall in this category are the IATF 16949 for the automotive industry and Q1 specifically for Ford Motor Co. There are many more for other industries as well.

This book is about how to validate the prescribed "quality" called out in the Quality Manual (if it exists), procedures, instructions, and standards (whether international, industrial, or organizational). It addresses the structure and the requirements of acceptable documentation for a successful validation of a given organization's quality. It discusses a detailed approach to fulfilling the requirements of the documentation process to the international standards as well as some of the automotive industry and specific requirements of the three largest car companies. These are emphasized by explaining the role of the auditee and auditor.

As such, the book has two basic objectives. The first objective is to provide a reference and a manual for the generalist who deals with documentation. As a result, the target audience is anyone who is interested in or involved with documenting quality issues. The second objective is to explain the methods and tools of the auditing process to help an organization gain and maintain the certification. Thus, the target audience is the individual(s) in any organization who is responsible for the implementation of standards or requirements to "fulfill" customer "requirements, wants, and expectations."

Acknowledgments

As any endeavor in life, it takes more than one person to complete a task. The task for writing this book is no different. I have consulted, interviewed, and have had discussions with many professionals in the auditing industry, and I am indebted to so many for their contribution. Without their shared information, without their enthusiasm and prodding me to finish, this work would not have been completed.

I feel very bad that I am not able to give credit to all who helped in the way of completing this book; however, I want everyone who did help to know that it is greatly appreciated, and I owe you a very big thank you. In that long list of the unwritten thanks, I do want to mention some individuals who helped tremendously in my effort to organize, design, and present this book for publication. They are

- Jessica Stamatis (my daughter-in-law) for helping with the artwork and the general computer work every time I got stuck with some triviality. She was always there to help with a smile. Thanks, Jess.
- Jerome Nosal who was available anytime to discuss customer specifications and, in many instances, to clarify some of the requirements. Thanks, Jerome!
- I am indebted to Pretesh Biswas, APB Consultant at http://isoconsultantpune. com/apb-consultant-iso-90012015-quality-management-system/apb-consultant-seven-principles-quality-management-per-iso-90012015/ for his permission to use them. His source was Art Lewis and the website of advisera.com and many others.
- Timothy and Cary Stamatis (my sons) who at the "drop of the hat" were available for discussing industrial standards and computer issues, respectively. Thanks guys. I really appreciated your effort.
- Thanks to a very close friend Anthony Roark for his relentless availability and willingness to help this project. As a professional auditor for 20+ years, his comments, suggestions, and overall editing were unbelievably helpful. Anthony passed on May 4, 2020, but he remains in our hearts. May your memory be eternal good friend.
- NSF's list provided on their website is used with permission: http://www. nsf.org/newsroom_pdf/isr_changes_iso14001.pdf. Retrieved on November 24, 2019.
- Finally, I want to thank my wife Carla for her relentless support and patience that she has showed me over the years. Her availability to answer questions about clarity, editing, and proper order of thoughts has made this book much better than it would have been without her input. Thank you so much Cara Jeanne.

Author

D. H. Stamatis, Ph.D., CQE, CMfgE, MSSBB, is the president of Contemporary Consultants, specializing in management, organizational development, and quality science applications. He has more than 40 years of experience in the academic world (teaching undergraduate- and graduate-level courses in statistics, operational management, project management, cultural management, supplier quality, and economics) and in industry (training and consulting worldwide in a variety of industries and organizations such as automotive, defense, electronic, health care, medical devices, financial institutions, and printing). Stamatis received his BS and BA degrees from Wayne State University, MA from Central Michigan University, and Ph.D. from Wayne State University. He is also an American Society of Quality (ASQ) fellow.

Introduction

Everyone wants excellent product and/or service. However, the way we define excellence is problematic since not everyone has the same definition. For our purposes, excellence in quality is defined by the customer. As such, organizations find themselves in a predicament that forces them to have some sort of standardization.

For the majority of quality issues, concerns, and problems that are encountered in many industries, we have some sort of standardization. However, this standardization is found in the international (ISO) and industry standards (IATF 16949), but it is not 100% accountable for all idiosyncrasies of many organization. This brings us to the customer-specific requirements. Each organization has its own variation of a particular characteristic of quality, and they publish them accordingly.

In the automotive industry, all three standards/requirements/specifications (ISO, IATF, individual specifications) are needed to do business with the OEMs (FCA, GM, Ford, etc.). To be sure, standardization has been instrumental in total improvement especially in our fast-paced movement towards zero defects and 100% customer satisfaction. Standardization has helped organizations to

- Define a quality system that is appropriate and applicable to a given organization
- Demonstrate the organization's (management's) commitment to a management system that maintains quality to their customers
- Compete in international markets
- Follow standard safety and product liability regulation and procedures
- Reduce cost and provide a practical results-oriented target(s)
- Help themselves maintain quality improvement gains
- Minimize supplier surveillance – through second-party audits
- Provide a platform from which to launch a continual improvement program such as total quality management (TQM) or some other program (Shingo, Malcolm Baldrige, etc.).
- Involve ALL employees by stimulating understanding of quality systems and their effect on the organization and its customers.

To help us identify "gaps" between what *we want* and what *we get*, we use audits to evaluate our system and to introduce remedies that will either close the gap or eliminate it completely. Audits however, if they are to be conducted with the seriousness they deserve, are time-consuming and costly. Therefore, all audits, without exception, require preparation. This preparation entails planning. This means having knowledge of the process being audited (training), knowledge of what is expected as an outcome and how to do the actual audit planning.

To facilitate these prerequisites, a checklist is strongly recommended. It is a path (map) of what, where, how, and why things are completed in a particular way, and it guides (facilitates) the flow of information to document whether or not all the requirements are being fulfilled. If not, you pursue deeper and find out what is missing, why is missing, and what can you do about it. In some cases, a checklist is a way to make sure you do not forget significant questions but more importantly will help the auditor to identify "trails" that were not originally thought of.

In this book, we provide the reader with an overall review of the standards and several possible checklists that can help in the process of auditing. We have included sample checklist for the ISO standards, industry standards, and customer-specific standards. We are aware that in all cases, predefined checklists have been published by all the entities involved; however, for many suppliers, these documents may be difficult to obtain and/or there are many suppliers which on their own want to improve through audits. This book hopefully will help in pursuing continual improvement. Specifically, the book covers the following:

> *Chapter 1*: Introduction to international quality standards: It provides an overview of the standards and their requirements.
>
> *Chapter 2*: Introduction to industry quality standards: It provides an overview of the automotive industry standards and their requirements.
>
> *Chapter 3*: Introduction to customer-specific requirements: It provides an overview and discussion of the organizational standards for Ford's, Fiat-Chrysler's, and General Motors' specific requirements.
>
> *Chapter 4*: Documentation: It discusses the documentation process and the requirements.
>
> *Chapter 5*: Checklist: It provides a rationale and several examples of checklists for the ISO standards, IATF, and customer-specific requirements.

1 Introduction to International Quality Standards

OVERVIEW

ISO is an independent, non-governmental international organization with a membership of 164 national standards bodies. Its origin began in 1926 as the International Federation of the National Standardizing Associations (ISA). This organization focused heavily on mechanical engineering. It was disbanded in 1942 during the Second World War but was re-organized in October 1946 with delegates from 25 countries under the name of United Nations Standards Coordinating Committee (UNSCC) with a proposal to form a new global standards body. The new organization officially began operations in February 1947, having the central secretariat at Geneva, Switzerland. The main roles and responsibilities of ISO are just to draft and publish standards (https://www.iso.org/standards.html. Retrieved on January 10, 2020).

As useful as the ISO standards are, there is a confusion about its name. Contrary to public usage and knowledge, ISO IS NOT an acronym. Rather, it is the Greek word ίσος, meaning "equal." This was selected purposely since if they had chosen the name of the country, there would have been different acronyms as the words would have been different. For example, the name of the organization in French is *Organisation internationale de normalisation* and, in Russian, Международная организация по стандартизации (*Mezhdunarodnaya organizatsiya po standartizatsii*). One can appreciate the ISO selection as the official name for its clarity, simplicity, and ease of usage worldwide. In the words of the ISO organization, we read the explanation thusly: "Because 'International Organization for Standardization' would have different acronyms in different languages (IOS in English, OIN in French), our founders decided to give it the short form *ISO*. *ISO* is derived from the Greek *isos*, meaning equal. Whatever the country, whatever the language, the short form of our name is always *ISO*." As such both the name *ISO* and the ISO logo are registered trademarks, and their use is restricted (about us: https://www.iso.org/standards.html. Retrieved on January 10, 2020).

Through its members, it brings together experts to share knowledge and develop voluntary, consensus-based, market-relevant international standards that support innovation and provide solutions to global challenges.

International standards make things work. They give world-class specifications for products, services, and systems, to ensure quality, safety, and efficiency. They are instrumental in facilitating international trade through standardization. As of July 2019, the ISO has published 22,933 standards and related documents, covering almost every industry, from technology, to food safety, to agriculture and healthcare. ISO international standards impact everyone, everywhere. Even though the standards have been translated in many languages, there are three official languages: English (with Oxford spelling), French, and Russian (https://www.iso.org/standards.html. Retrieved on January 10, 2020).

The most popular of the standards are the 9000 (quality), 14000 (environmental), and 45000 (occupational, health, and safety) series. All of them are reviewed every 5 years and revised as necessary. All of them are audited for conformance and/or compliance depending on the standard. Other standards also cover day-to-day activities that affect us all, including cinematography, shoes sizes, thermal insulation, and textiles.

The breakdown of the count of ISOs produced thus far are

- International Standard = 20,038
- Technical Report = 849
- Technical Specifications = 559
- Guide = 40
- Publicly Available Specifications = 28
- International Standardized Profile = 13
- International Workshop Agreement = 11
- Technology Trends Assessment = 5
- Recommendation = 1.

In this book, we will focus only on the three major ISO standards that deal with quality primarily in the automotive industry. We believe that these standards, when used appropriately, help the organization that uses them in creating products and services that are safe, reliable, and of good quality. The standards help businesses increase productivity while minimizing errors and waste. By enabling products from different markets to be directly compared, they facilitate companies in entering new markets and assist in the development of global trade on a fair basis. The standards also serve to safeguard consumers and the end-users of products and services, ensuring that certified products conform to the minimum standards set internationally (https://www.iso.org/standards.html. Retrieved on January 10, 2020).

When one discusses quality in any form, invariably the topic will be something like: But what standards are being followed? Or what quality management standards (QMS) the organization operates under?

To answer these fundamental questions, let us examine them in a very cursory form. First, the issue of QMS and then the practical meaning of "quality standards" in very general terms. So, a quality management standard establishes a skeleton for how an organization manages its key activities in a given enterprise. This, of course, is an agreement within the organization that provides the "way" of doing something, either making a product, managing a process, or delivering a service. On the other hand, a quality standard is a detail document detailing the requirements, specifications, guidelines, and characteristics that products, services, and processes should consistently meet in order to ensure specific characteristics such as

- Their quality matches expectations.
- They are fit for purpose.
- They meet the needs of their customers and ultimate users.
- Some form of standards is an essential element of any QMS.

Therefore, one may come to a conclusion that the purpose of quality management standards is simply a way for businesses to satisfy their customers' quality requirements and for a range of other reasons, such as

- Ensuring safety and reliability of their products and services
- Complying with regulations, often at a lower cost
- Defining and controlling internal processes
- Meeting environmental objectives
- Meeting customer's needs and expectations.

When an organization is "truly" committed to following quality management standards, they are often more likely to

- Increase their profits
- Reduce losses or costs across the business (through less rejects and rework)
- Improve their competitiveness
- Gain market access across the world
- Increase consumer loyalty (due to consistent quality).

The international standards are one way to demonstrate commitment to quality and validate that quality with audits to a given organization. To be sure, ALL international standards fall under the control of the International Organization for Standardization (ISO), which is based in Zurich, Switzerland. In this book, we will address ONLY some of the common ones that deal with quality, environment, occupational hazards and safety as well as the automotive standard known as the IATF 16949 and three customer requirements from Ford Motor Company, General Motors, and Fiat Chrysler Automotive.

ISO 9001

General Comments

This standard is considered to be the basic generic quality standard for ALL organizations that are interested in quality. Over the years, it has been revised several times, and the current one is the ISO 9000:2015 series which provides the structure for the entire standard. However, the certifiable standard is the ISO 9001:2015 which covers the seven specific principles of any QMS. They are

1. *Customer focus*: The primary focus of quality management is to meet customer requirements and to strive to exceed customer expectations.
2. *Leadership*: Leaders at all levels establish unity of purpose and direction and create conditions in which people are engaged in achieving the quality objectives of the organization.
3. *Engagement of people*: It is essential for the organization that all people are competent, empowered, and engaged in delivering value. Competent, empowered, and engaged people throughout the organization enhance its capability to create value.
4. *Process approach*: Consistent and predictable results are achieved more effectively and efficiently when activities are understood and managed as interrelated processes that function as a coherent system.
5. *Improvement*: Successful organizations have an ongoing focus on improvement.
6. *Evidence-based decision-making*: Decisions based on the analysis and evaluation of data and information are more likely to produce desired results.
7. *Relationship management*: For sustained success, organizations manage their relationships with interested parties, such as suppliers.

The key changes from previous revisions to the new ISO 9001:2015 which must be addressed in any audit dealing with this standard are as follows:

1. There is no requirement for a quality manual (however, highly recommended).
2. Its emphasis on organizational context and risk-based thinking is of paramount importance.
3. There is no requirement for the management representative. Now the leadership clause is more inclusive for the top management.
4. The standard does not include a specific clause for "Preventive Actions." Now preventive actions are sprinkled in the entire standard.
5. The terms "document" and "records" have been replaced with the term "documented information." Documented procedure in ISO 9001:2008 has been replaced by maintained documented information, and documented record in ISO 9001:2008 has been replaced by retained documented information. However, for all intents and purposes, the requirements are the same as before. The names have changed.
6. In 2008 version of the standard, the term "product" was used. Now it is more inclusive and covers both *Product* and *Services*.
7. The term "continual improvement" has been enriched with the addition of the term "improvement."
8. Outsourcing is now an external provider. The term "purchased product" has been replaced with "externally provided products and services." The term "supplier" has been replaced with "external provider." Control of external provision of goods and services addresses all forms of external provisions.
9. The new standard does not make any reference to the exclusions which was only for clause 7 in ISO 9001:2008. However, in ISO 9001:2015 after proper

justification any of the requirement of this international standards may not be included in the scope, provided that it does not affect the organization's ability or responsibility to ensure the conformity of its product and services and the enhancement of customer satisfaction.

10. The term "work environment" now is replaced with "environment for the operation of processes."

Among the many benefits of this generic quality standard are as follows:

- There is an increase in customer value.
- There is an increase in customer satisfaction.
- There is an improvement in customer loyalty.
- There is noticeable enhancement in repeat business.
- There is an increase in the reputation of the organization.
- There is an expansion of the customer base.
- There is an increase in revenue and market share.
- There is an increase in the effectiveness and efficiency in meeting the organization's quality objectives.
- There is better coordination of the organization's processes.
- There is an improvement in communication between levels and functions of the organization.
- There is a propensity to develop and improve the capability of the organization and its people to deliver the desired results.
- There is an improvement in understanding of the organization's quality objectives by people in the organization and increased motivation to achieve them.
- There is an increase in involvement of people in improvement activities.
- There is an increase in personal development, initiatives, and creativity.
- There is an increase in people's satisfaction.
- There is an increase in trust and collaboration throughout the organization.
- There is an increase in attention to shared values and culture throughout the organization.
- There is an increase in the ability to focus effort on key processes and opportunities for improvement.
- There are consistent and predictable outcomes through a system of aligned processes.
- There is an increase and willingness to optimize performance through effective process management, efficient use of resources, and reduced cross-functional barriers.
- There is a willingness to enable the organization's leadership to provide confidence to interested parties related to its consistency, effectiveness, and efficiency.
- There is improved process performance, organizational capability, and customer satisfaction.
- There is an enhanced focus on root cause investigation and determination, followed by prevention and corrective actions.

- There is an enhanced ability to anticipate and react to internal and external risks and opportunities.
- There is enhanced consideration of both incremental and breakthrough improvement;
- There is improved use of learning for improvement. There is an enhanced drive for innovation.
- There is an improvement in decision-making processes.
- There is an improvement in the assessment of process performance and ability to achieve objectives.
- There is an improvement in operational effectiveness and efficiency.
- There is an increased ability to review, challenge, and change opinions and decisions;
- There is an increased ability to demonstrate the effectiveness of past decisions.
- There is enhanced performance of the organization and its relevant interested parties through responding to the opportunities and constraints related to each interested party.
- There is a common understanding of objectives and values among interested parties.
- There is an increased capability to create value for interested parties by sharing resources and competence and managing quality-related risks.
- There is a well-managed supply chain that provides a stable flow of products and services.

The actions that one may take to conform to validate these requirements and benefit from them are many. However, these are some essentials that both the auditee and auditors must – at least – consider to incorporate into their checklist to pursue and validate the QMS.

- It must identify and recognize the direct and indirect customer of the organization who receive value from the organization.
- It must understand customers' current and future needs and expectations.
- It must link the organization to its objectives, to customer needs and expectations.
- It must communicate customer needs and expectations throughout the organization.
- It must plan, design, develop, produce, deliver, and support products and services to meet customer needs and expectations.
- It must measure and monitor customer satisfaction and take appropriate actions.
- It must determine and take action on relevant interested parties' needs and appropriate expectations that can affect customer satisfaction.
- It must actively manage relationships with customers to achieve sustained success.

- It must communicate the organization's mission, vision, strategy, policies, and processes throughout the organization.
- It must create and sustain shared values, fairness, and ethical models for behavior at all levels of the organization.
- It must establish a culture of trust and integrity.
- It must encourage an organization-wide commitment to quality.
- It must ensure that leaders at all levels are positive examples to people in the organization.
- It must provide people with the required resources, training, and authority to act with accountability
- It must inspire, encourage, and recognize the contribution of people.
- It must communicate with people to promote understanding of the importance of their individual contribution.
- It must promote collaboration throughout the organization.
- It must facilitate open discussion and sharing of knowledge and experience.
- It must empower people to determine constraints to performance and to take initiatives without fear.
- It must recognize and acknowledge people's contribution, learning, and improvement.
- It must enable self-evaluation of performance against personal objectives.
- It must conduct surveys to assess people's satisfaction, communicate the results, and take appropriate actions.
- It must define the objectives of the system and processes necessary to achieve them.
- It must establish authority, responsibility, and accountability for managing processes.
- It must understand the organization's capabilities and determine resource constraints prior to action.
- It must determine process interdependencies and analyze the effect of modifications to individual processes on the system as a whole.
- It must manage processes and their interrelations as a system to achieve the organization's quality objectives effectively and efficiently.
- It must ensure the necessary information is available to operate and improve the processes and to monitor, analyze, and evaluate the performance of the overall system.
- It must manage risks which can affect outputs of the processes and overall outcomes of the QMS.
- It must determine, measure, and monitor key indicators to demonstrate the organization's performance.
- It must make all data needed available to the relevant people.
- It must ensure that data and information are sufficiently accurate, reliable, and secure.
- It must analyze and evaluate data and information using suitable methods.
- It must ensure people are competent to analyze and evaluate data as needed.

- It must make decisions and take actions based on evidence, balanced with experience and intuition.
- It must determine relevant interested parties (such as providers, partners, customers, investors, employees, or society as a whole) and their relationship with the organization.
- It must determine and prioritize interested party relationships that need to be managed.
- It must establish relationships that balance short-term gains with long-term considerations.
- It must (and can) gather and share information, expertise, and resources with relevant interested parties.
- It must (and can) measure performance and provide performance feedback to interested parties, as appropriate, to enhance improvement initiatives.
- It must establish collaborative development and improvement activities with providers (suppliers), partners, and other interested parties;
- It must encourage and recognize improvements and achievements by providers and partners.

Some of the above questions and statements are a combination of several sources. I am indebted to Pretesh Biswas, APB Consultant at http://isoconsultantpune. com/apb-consultant-iso-90012015-quality-management-system/apb-consultant-seven-principles-quality-management-per-iso-90012015/ for his permission to use them. His source was Art Lewis and the website of advisera.com, and many others.

ISO 9000/9001

ISO 9000:2015 describes the fundamental concepts and principles of quality management which are universally applicable to the following:

- Organizations seeking sustained success through the implementation of a quality management system.
- Customers seeking confidence in an organization's ability to consistently provide products and services conforming to their requirements.
- Organizations seeking confidence in their supply chain that their product and service requirements will be met.
- Organizations and interested parties seeking to improve communication through a common understanding of the vocabulary used in quality management.
- Organizations performing conformity assessments against the requirements of ISO 900.

ISO 9000:2015 specifies the terms and definitions that apply to all quality management and quality management system standards developed by ISO/TC 176. However, it is not a certifiable standard. Certification is reserved ONLY for ISO 9001.

Specifically, the ISO 9001 standard specifies requirements for a quality management system (QMS) which is based on the quality management principles described in ISO 9000. Organizations use the standard to demonstrate the ability to consistently provide products and services that meet customer and regulatory requirements. It is the most popular standard in the ISO 9000 series and the only standard in the series to which organizations can certify.

The ISO 9001:2015 quality standard is a set of requirements that affect virtually all aspects of the operations of corporate enterprises, non-profit organizations, and government entities. ISO 9001 sets the requirements for a company's Quality Management System (QMS), but a QMS is very different from quality control. Quality control refers to checking product or service features to verify they conform to requirements, while a Quality Management System is about the management of the entire enterprise and its operational processes. [That is why an organization may have a good QMS but not so good product or service quality.]

ISO 9001 is designed for any company (in fact, for any organization) of any size and in any industry. This broad application results in ISO 9001 standard being rather broad and its requirements rather difficult to read and understand. The following is a summary of the ISO 9001:2015 requirements in lay man's terms.

Clauses 0–3: Introductory Chapters

Clauses 0–3 of ISO 9001:2015 are introductory chapters that don't contain any requirements and therefore are not certifiable. This particular structure is followed in ALL ISO standards.

Introduction

1. Scope of ISO 9001
2. Normative references
3. Terms and definitions
4. Context of the organization.

The introduction of these categories includes information on the quality management principles and the process approach, both of which form the basis of ISO 9001:2015. There is also some information on how ISO 9001:2015 relates to other ISO standards. The current version of ISO 9001 was released in September 2015. The following are the quality management principles that apply to all organizations (manufacturing as well as service – regardless of size or industry):

1. Customer focus
2. Leadership
3. Engagement of people
4. Process approach
5. Improvement
6. Evidence-based decision-making
7. Relationship management.

The focus of the standard is to make sure that

- The processes of the organization are organized effectively and documented.
- The identification of processes that need improvement in efficiency is identified.
- The identification of processes that need "continual improvement" is actively pursued.

To do that, the standard utilizes the classic model of the *plan–do–check (study)–act* methodology and provides a process-oriented approach to documenting and reviewing the structure, responsibilities, and procedures required to achieve effective quality management in an organization. Specific sections of the standard contain information on many topics, such as

- Requirements for a quality management system, including documented information, planning, and determining process interactions
- Responsibilities of management
- Management of resources, including human resources and an organization's work environment
- Product realization, including the steps from design to delivery
- Measurement, analysis, and improvement of the QMS through activities like internal audits and corrective and preventive action.

The new ISO 9001 revision – although it kept the past revisions – is again updated to reflect the continuing changing environments in which organizations operate. Some of the key updates include

- The introduction of new terminology
- Restructuring some of the information
- An emphasis on risk-based thinking to enhance the application of the process approach
- Improved applicability for services
- Increased leadership requirements.

So, why should an organization pursue the ISO 9001 certification? Are there any benefits to it? If the management of an organization wants to ensure that their customers consistently receive high quality products and services, which in turn brings many benefits, including satisfied customers, management, and employees, then the answer is categorically YES, there are tangible benefits worth pursuing. There is no doubt that ISO 9001 specifies the requirements for an effective quality management system, and organizations find that using the standard helps them

- Organize a quality management system (QMS)
- Create satisfied customers, management, and employees

- Continually improve their processes
- Save costs.

Certification

As already mentioned, to be able to claim all the benefits of following the structure of the ISO 9001 the organization must be certified. In fact, some customers demand that their supplier base be certified. So, certification means that an organization has demonstrated the following:

- Follows the guidelines of the ISO 9001 standard
- Fulfills its own requirements
- Meets customer requirements and statutory and regulatory requirements
- Maintains documentation.

The certification process includes implementing the requirements of ISO 9001:2015 (evidence of an internal audit) and then completing a successful registrar's audit (third-party audit) confirming the organization meets those requirements.

Contents of the ISO 9001:2015 Fifth ed.

The ISO 9001:2015 quality standard is a set of requirements that affect virtually all aspects of the operations of corporate enterprises, non-profit organizations, and government entities.

ISO 9001 sets the requirements for a company's Quality Management System (QMS), but a QMS is very different from quality control. Quality control refers to checking product or service features to verify they conform to requirements, while a Quality Management System is about the management of the entire enterprise and its operational processes.

ISO 9001 is designed for any company (in fact, for any organization) of any size and in any industry. This broad application results in ISO 9001 standard being rather broad and its requirements rather difficult to read and understand. The following is a summary of the ISO 9001:2015 requirements in lay man's terms.

Clauses 0–3: Introductory Chapters – Non-certifiable

Clauses 0–3 of ISO 9001:2015 are introductory chapters that don't contain any requirements.

Introduction

1. Scope of ISO 9001
2. Normative references
3. Terms and definitions.

The introduction includes information on the quality management principles and the process approach, both of which form the basis of ISO 9001:2015. There is also some information on how ISO 9001:2015 relates to other ISO standards.

Clause 4: Context of the Organization

4.1 Understanding the organization and its context
4.2 Understanding the needs and expectations of interested parties
4.3 Determining the scope of the quality management system
4.4 Quality management system and its processes

This section sets the requirements for the foundation of the ISO 9001 Quality Management System. The "context of the organization" is the business environment in which the company operates. The first requirement is to identify external and internal *strengths and weaknesses that* are relevant to the company and the ISO 9001 Quality Management System. Next, the *needs and expectations* of not only the customers but a wide range of "interested parties" (or stakeholders) need to be determined, and how they are relevant to the ISO 9001 Quality Management System. Here you also need to decide on the *scope* of the ISO 9001 Quality Management System (and the ISO 9001 certification). While it is possible to exclude functions and certain products or services, the scope will in most cases be the entire company.

Finally, you will need to use the *process approach* to determine the processes of the ISO 9001 Quality Management System. The Quality Management System will need to be implemented, maintained, and continually improved upon. The *documentation* of the Quality Management System must include procedures and work instructions to ensure the effective operation and control of all processes. This documentation must be controlled to ensure the correct people have access to current revisions. *Records* are established and controlled to provide evidence of a properly maintained ISO 9001 Quality Management System. Remember that "Quality Management System" refers to integrated management processes throughout the entire company. [A very important reminder here is the misconception that the ISO 9001:2015 requires less documentation than previous revisions of the standard because there is no longer an explicit requirement for a quality manual. However, ISO 9001:2015 contains numerous implicit documentation requirements. Since most companies consider the documentation the most difficult part of the ISO 9001 implementation, the use of good ISO 9001 templates can be beneficial.]

Clause 5: Leadership

5.1 Leadership and commitment
    5.1.1 General
    5.1.2 Customer focus
5.2 Policy
    5.2.1 Establishing the quality policy
    5.2.2 Communicating the quality policy
5.3 Organizational roles, responsibilities, and authorities

This section of ISO 9001:2015 is all about the involvement of top management in the ISO 9001 Quality Management System. The first part of this clause summarizes the various responsibilities of top management with regard to the ISO 9001

Quality Management System. Among those is the requirement that top management *integrate the ISO 9001 Quality Management System into the operational processes* of the company, and aligns policy and objectives with the strategy of the company. Here, top management must take leadership when it comes to *customer focus*, including determining customer requirements, determining related risks and addressing them, and maintaining a focus on customer satisfaction. Furthermore, top management establishes its *quality policy*. Special attention is given to a commitment to meeting requirements and to continual improvement, and a framework for establishing and reviewing quality objectives. The focus here is to ensure that *responsibilities and authorities* within the company are clearly established. A task or job can only be accomplished if it is clear who is responsible for it. Ultimately, top management is responsible for the ISO 9001 Quality Management System, but they may appoint an *ISO 9001 Management Representative* who takes on various responsibilities and authorities of the ISO 9001 Quality Management System. [Remember that the use of simple documentation (both text and forms) will facilitate the conformance and/or compliance to the standard as applicable and appropriate.]

Clause 6: Planning

 6.1 Actions to address risks and opportunities
 6.2 Quality objectives and planning to achieve them
 6.3 Planning of changes

This section focuses on planning and the concept of risk-based thinking. The first part addresses how the company needs to engage in *risk and opportunity management*. ISO 9001 doesn't dictate action to address all risks and opportunities but requires to set up a system of evaluating risks and opportunities in order to make an informed decision as to what (if any) action is needed. Second, senior management needs to establish measurable *quality objectives* and plan how to achieve them. Those quality objectives are basically strategic objectives of the company that are relevant to product and service requirements, as well as customer satisfaction. Finally, one should pay attention to the 6.3 element which addresses the *planning of changes*. This must be addressed in a systematic manner and NOT brushed over. [Remember that risk management does not need to be complicated or require the establishment of a new department or the hiring of a risk manager. Small to mid-size companies may choose to use a risk management matrix and calculate a priority number – it's a simple, yet effective system.]

Clause 7: Support

 7.1 Resources
  7.1.1 General
  7.1.2 People
  7.1.3 Infrastructure
  7.1.4 Environment for the operation of processes
  7.1.5 Monitoring and measuring resources
  7.1.6 Organizational knowledge

7.2 Competence
7.3 Awareness
7.4 Communication
7.5 Documented information
    7.5.1 General
    7.5.2 Creating and updating
    7.5.3 Control of documented information

This section of ISO 9001:2015 is all about support functions: various resources, competence and training, communication, and documentation. The first part of this clause clarifies the requirement for a company to determine and provide, in a timely manner, *resources* needed to implement and improve the processes of the Quality Management System. Resources include people (*human resources*) and their competence, as well as training necessary to achieve the required competencies. As we continue to read this section, we are expected to identify, provide, and maintain the *infrastructure* needed to achieve the conformity of its products or services. Infrastructure includes buildings, equipment, transportation, and communications technology. The company must identify and manage those human and physical factors of the *work environment* needed to achieve conformity of its products or services.

- *Maintenance* activities of equipment, machinery as well as the work environment are part of this section. The company must take special care of all *measuring devices* and properly calibrate them.
- *Organizational knowledge* is another resource that must be determined, maintained, and, importantly, shared so that the company can fully utilize it. Organizational knowledge is a wide-ranging term, which includes best practices. When addressing the needed *competence* of its people, the company must not only consider training but also hiring and reassignment of people.
- *Awareness* is another part of clause 7. Management and staff must be aware of the quality policy and relevant objectives, as well as their part in the ISO 9001 Quality Management System. Further, the company needs to have established internal and *external communications channels*.
- Finally, element 7 addresses *documents and records*. Essentially, the company must control its documents to ensure that the right people have the current version of the right document available. Records must be kept for numerous activities. [Remember integration is important for this section of the standard. Therefore, integrate your company's HR function well into your ISO 9001 quality system and make them take on a leading role during the ISO 9001 implementation.]

Clause 8: Operation

8.1 Operational planning and control
8.2 Requirements for products and services

8.2.1 Customer communication

8.2.2 Determining the requirements for products and services

8.2.3 Review of the requirements for products and services

8.2.4 Changes to requirements for products and services

8.3 Design and development of products and services

  8.3.1 General

  8.3.2 Design and development planning

  8.3.3 Design and development inputs

  8.3.4 Design and development controls

  8.3.5 Design and development outputs

  8.3.6 Design and development changes

8.4 Control of externally provided processes, products, and services

  8.4.1 General

  8.4.2 Type and extent of control

  8.4.3 Information for external providers

8.5 Production and service provision

  8.5.1 Control of production and service provision

  8.5.2 Identification and traceability

  8.5.3 Property belonging to customers or external providers

  8.5.4 Preservation

  8.5.5 Post-delivery activities

  8.5.6 Control of changes

8.6 Release of products and services

8.7 Control of nonconforming outputs

In this section, the ISO 9001:2015 standard sets the requirements for the processes needed to achieve the product or service. This is how your product or service is designed, produced, tested, handled, shipped, etc. Emphasis is placed on how the company understands, communicates, and meets *customer requirements*, and what it does if customer requirements change. *Design* and development reviews, verification, and validation must be planned for at the very beginning of the design and development process. This section also addresses the *purchasing* of products and services and *outsourcing*. Controls include supplier evaluations, selection, disqualification, and receiving inspections. The element specifies several controls for the actual *production and service provision*, ranging from work instructions to QC inspections. There are also requirements for the identification of components and the ability of tracing a product or service back, as well as for handling products that belong to customers or suppliers. The standard also addresses how the final product or service can eventually be *released* to the customer. Finally, this element addresses cases in which output is found as *not conforming* to requirements, including cases in which such a nonconformance is only detected after the product has been delivered or the service provided. [Remember that written work instructions and flowcharts not only define but also force standardization of understanding processes. Due to the nature of standardization, much time can be saved if you follow the ISO 9001 requirements for document control from the beginning.]

Clause 9: Performance Evaluation
9.1 Monitoring, measurement, analysis, and evaluation
    9.1.1 General
    9.1.2 Customer satisfaction
    9.1.3 Analysis and evaluation
9.2 Internal audit
9.3 Management review
    9.3.1 General
    9.3.2 Management review inputs
    9.3.3 Management review outputs

This section is all about measuring and evaluating. *Measurement and monitoring activities* needed to assure conformity and achieve improvement must be defined, planned, and implemented. Measuring and monitoring allows the company to manage by fact, not by guess. The company must monitor information relating to *customer perception* as to whether the company has fulfilled customer requirements. Furthermore, this section focuses on the *analysis of data* arising from monitoring and measurement activities. There are several methods to analyze data, including statistical techniques, but the outcome of the analysis is designed to gain the necessary information to make fact-based (data-driven) decisions.

Internal auditing of the company must be done at planned intervals to understand if the ISO 9001 Quality Management System is working as planned. Audits are checks on the system, not on individuals. Finally, *management reviews* will need to be carried out. The management reviews cover a wide range of topics related to the ISO 9001 Quality Management System, ranging from customer satisfaction to the performance of suppliers. The results of management reviews are decisions and actions regarding improvements, changes and resource needs. [Remember that an effective audit is based on the results of the internal audit. Therefore, the earlier the internal audit is completed, the better the probability of finding and fixing problems or potential problems in the system and fixed before the certification (third-party audit) be conducted.]

Clause 10: Improvement
10.1 General
10.2 Nonconformity and corrective action
10.3 Continual improvement

The last section of ISO 9001:2015 requires companies to determine and identify opportunities for improvement. There is a requirement for the *improvement* of products and services with an eye to current and future market needs. This element also includes requirements regarding the correction of nonconformities. First, companies must react to the *nonconformity*. Second, companies must engage in *corrective action* to correct the underlying causes of existing problems, as well as engage in preventive action to correct situations that could lead to potential problems. Finally, the company must plan and manage the *processes for the continual improvement* of the ISO 9001 Quality Management System. The company must use the quality policy, objectives, internal audit results, analysis of data, corrective and preventive action, and management review to facilitate continual improvement. [Remember that this final section of the standard is of paramount importance in that it establishes the effectiveness of the implementation process. If done appropriately and thoroughly, it can add much value. If not, it will end up with a large bureaucracy and waste. Some of the materials have been adopted from https://www.9001council.org/iso-9001-summary.php. January 13, 2017.]

So, what is the role of the auditor and the auditing process? Simply put the functions are different but supplemental to each other. First, the auditing process is focused on whether the QMS is covered based on the standard, and second, the auditor is responsible to identify the effectiveness of the QMS as defined by the organization. Therefore, based on the industry and the organization's QMS, the lead auditor will generate a checklist and then move into the physical audit, asking questions and verifying the documentation presented to them. In the final analysis, the auditor will determine the certification to the standard and whether or not conformances are documented with objective evidence.

ISO 9001 Documentation Requirements

The ISO 9001:2015 certification is as much about the documentation as it is the process of standardization. One could even say the two goals are one and the same. When an organization applies for certification, the application rests on the long list of documents and records that share your processes, procedures, and standards.

So, when the ISO 9001:2015 requirements launched, it changed the documentation requirements. The new version demands far less paperwork from businesses, which also means it's applicable to more companies. At the same time, it can also trip you up: you still need to document every process, procedure, and standard that you identify as necessary regardless of whether the standards explicitly name the documents. These are the non-mandatory requirements, but only in the sense that you don't need them if the process doesn't apply to you. *Essentially, the organization is allowed to determine for itself just how it prefers to document the QMS as well as what management believes is the adequate amount of documentation to represent its planning.*

To explore what this means in real terms, let's dive into the ISO 9001:2015 documentation requirements. To be sure, documented information is the Core of the ISO 9001:2015. This means that the paperwork – not the processes per se – is at the heart of *the* ISO 9001:2015 development journey. The standard refers to it as "documented information." Therefore, the documented information is the meaningful information and data that requires control and that the organization must maintain. The documents can refer to

- The Quality Management System (QMS) generally
- The processes
- The documentation
- The records.

Your documented information can communicate messages, serve as evidence of implementation, or share knowledge between parties. According to the ISO's guidance on requirements, the main objectives of the documented information include

- Communicating information
- Providing evidence of conformity
- Knowledge sharing
- Disseminating and preserving the organization's experiences.

It is important to recognize that there are two basic ISO 9001:2015 documentation requirements. So, when applying for your ISO 9001 certification, *you will submit two groups of documents to the external auditor (this by the way may be performed by a distance audit which, of course, is used to be called desk audit and conducted in the organization's facility)*:

1. The documentation named by the standard (as provided below)
2. The documentation you decide is required for your QMS.

The Mandatory Documents Required by ISO 9001:2015

1. Documented information to the extent necessary to have confidence that the processes are being carried out as planned (clause 4.4).
2. Evidence of fitness for the purpose of monitoring and measuring resources (clause 7.1.5.1).
3. Evidence of the basis used for calibration of the monitoring and measurement resources (when no international or national standards exist) (clause 7.1.5.2).
4. Evidence of competence of person(s) doing work under the control of the organization that affects the performance and effectiveness of the QMS (clause 7.2).
5. Results of the review and new requirements for the products and services (clause 8.2.3).
6. Records needed to demonstrate that design and development requirements have been met (clause 8.3.2).
7. Records on design and development inputs (clause 8.3.3).
8. Records of the activities of design and development controls (clause 8.3.4).
9. Records of design and development outputs (clause 8.3.5).
10. Design and development changes, including the results of the review and the authorization of the changes and necessary actions (clause 8.3.6).
11. Records of the evaluation, selection, monitoring of performance and re-evaluation of external providers and any actions arising from these activities (clause 8.4.1).
12. Evidence of the unique identification of the outputs when traceability is a requirement (clause 8.5.2).

13. Records of the property of the customer or external provider that is lost, damaged, or otherwise found to be unsuitable for use and of its communication to the owner (clause 8.5.3).
14. Results of the review of changes for production or service provision, the persons authorizing the change, and necessary actions taken (clause 8.5.6).
15. Records of the authorized release of products and services for delivery to the customer including acceptance criteria and traceability to the authorizing person(s) (clause 8.6).
16. Records of nonconformities, the actions taken, concessions obtained, and the identification of the authority deciding the action in respect of the nonconformity (clause 8.7).
17. Results of the evaluation of the performance and the effectiveness of the QMS (clause 9.1.1).
18. Evidence of the implementation of the audit program and the audit results (clause 9.2.2).
19. Evidence of the results of management reviews (clause 9.3.3).
20. Evidence of the nature of the nonconformities and any subsequent actions taken (clause 10.2.2).
21. Results of any corrective action (clause 10.2.2).

These documents may look like a long list of busy work, as they need to be retained as records of the results of your QMS. However, there is a silver lining for those who think this form of documentation is excessive and burdensome. The good news is that many organizations already have these records as they exist in their current practice of their operation. What needs to be done – in some cases – is the reconfiguration of the existing documentation to align with the new requirements. Newer businesses (or those new to the ISO 9001:2015 standard) who don't have a long or broad documentation history are the ones who usually spend the most time generating the paperwork listed above.

Ultimately, the documented information is part of the core value of the ISO 9001:2015. It encourages you to standardize the processes you already employ and to work towards consistent data collection and data updates to core paperwork like the documents listed above.

A Guide to the Non-mandatory Documentation

In addition to the list of mandatory items, an organization as we mention earlier may choose to add documents that contribute to the value of the organization's QMS. These documents are called *non-mandatory*, and they include but are not limited to the following:

- Organization charts
- Procedures
- Specifications
- Work instructions
- Test instructions
- Production schedules

- Approved supplier lists
- Quality plans
- Strategic plans
- Quality manuals
- Internal communication documents.

If one does create these types of documented information, then one must follow the same rules laid out in clause 7.5. In other words, treat them the same way as you treat the named required documented information.

What about the quality manual? Those familiar with the ISO 9001:2008 will remember the need to complete and present the quality manual to the certification body before the audit.

The publication of the 2015 standard revealed that *the quality manual is no longer a mandatory document.*

If your organization received certification under the 2008 version, you might let out a cheer. But the truth is that even though it's no longer required, it is still a very helpful document to have. What's more, you still need some type of document that describes your QMS (including scope and process interactions) to send to the auditor, even if you don't call it a quality manual. Your organization's quality manual is a comprehensive look at your organization, your QMS, and the approach to quality management you selected on the development journey.

In the past, organizations spent a huge amount of time and energy creating these documents. While most met the fundamental requirements of the ISO 9001, many missed the spirit of the document itself. Instead, they focused too heavily on over-writing the document rather than creating something useful. If you created a quality manual in the past, there's a good chance that you never used it again.

So, even though it's no longer required, producing a quality manual can be incredibly helpful for your organization. It doesn't need to be 50+ pages: it should be short, snappy, and to the point with the goal of contributing to the QMS. Plus, you may find that some businesses require their suppliers to provide their quality manual during the tender or selection process. If you have a solid document to provide them, you'll set your organization apart from the competition.

How to Submit and Maintain Your Required Documentation

The ISO 9001:2015 standards aren't prescriptive in terms of the format of your documentation. You can submit paper documents, but you're also able to submit electronic versions of the documents. You can even use video and audio, if you choose. When setting up your documentation system, it is helpful to do so in a way that supports the maintenance of and retention of the documents. It should be easy to update, save, and share the documents when necessary. Choosing electronic documentation can help make the process of meeting clause 7.5 requirements, as dictated below.

When to Review Your ISO 9001:2015 Documentation

The ISO 9001:2015 requires that you control your documents, but it grants you much more freedom in doing so than the previous 2008 standard did. However, there are still requirements for updating the documented information.

Clause 7.5 requires you establish and use documented procedures to "maintain documented information to the extent necessary to support the operation of processes and retain documented information to the extent necessary to have confidence that the processes are being carried out as planned." For some businesses, the "extent necessary" may mean once a year. Others may only choose to go through the process once every 2 or 3 years. What is important here is to remember that when working through your obligation under clause 7.5, you need processes to

- Approve documents
- Review, update, and submit documents for re-approval
- Identify changes
- Make documents available
- Ensure documents are legible and identifiable
- Identify and control external documents
- Keep obsolete documents out of circulation
- Identify obsolete documents as necessary if retained.

You must also establish documented procedures for the following tasks:

- Identifying records
- Storing records
- Protecting records (including keeping them identifiable and legible)
- Retrieving records
- Retaining records
- Disposing of records.

Conclusion

The two most important objectives of the ISO 9001:2015 update is to develop a simpler set of standards that apply to all organizations and to allow organizations to focus on the most relevant documentation for their business activities. Therefore, as management of the organization it is very important to recognize most of the documented information from the list of requirements above. Your goal then is to standardize it and supplement it with the other documented information that will help your QMS work.

ISO 14000

GENERAL COMMENTS

ISO 14000 is similar to ISO 9000 quality management in that both pertain to the process of how a product is produced, rather than to the product itself. Specifically, however, ISO 14000 is a family of standards related to environmental management that exists to help organizations (a) minimize how their operations (processes, *etc.*) negatively affect the environment (i.e., cause adverse changes to air, water, or land), (b) comply with applicable laws, regulations, and other environmentally oriented requirements, and (c) continually improve in the above. Furthermore, ISO 14001:2015 sets out the criteria for an environmental management system (EMS). It does not state requirements for environmental performance, but maps out a framework that a company or organization can follow to set up an effective EMS.

It can be used by any organization that wants to improve resource efficiency, reduce waste, and drive down costs. Using ISO 14001:2015 can provide assurance to company management and employees as well as external stakeholders that environmental impact is being measured and improved. ISO 14001 can also be integrated with other management functions and assists companies in meeting their environmental and economic goals. ISO 14001 is voluntary, with its main aim to assist companies in continually improving their environmental performance, while complying with any applicable legislation. Organizations are responsible for setting their own targets and performance measures, with the standard serving to assist them in meeting objectives and goals and in the subsequent monitoring and measurement of these. The standard can be applied to a variety of levels in the business, from organizational level, right down to the product and service level. Rather than focusing on exact measures and goals of environmental performance, the standard highlights what an organization needs to do to meet these goals. ISO 14001 is known as a generic management system standard, meaning that it is relevant to any organization seeking to improve and manage resources more effectively. This includes

- Single site to large multi-national companies
- High-risk companies to low-risk service organizations
- Manufacturing, process, and the service industries, including local governments
- All industry sectors including public and private sectors
- Original equipment manufacturers and their suppliers.

Plan–Do–Study (Check)–Act Methodology of EMS

One of the most used methodologies in implementing and understanding the EMS is the application of the Deming/Shewhart model. It is shown in Table 1.1.

TABLE 1.1
The PDS(C)A Model

		Plan Clause 6		
Internal and external issues (4.1)				Support and operation (7 & 8)
	Act	Leadership (5)	**Do**	
Needs and expectations of interested parties (4.2)				Performance evaluation (9)
		Study (check) improvement (10)		

---

1. *Plan* – establish objectives and processes required – See clause 6. Just like in the implementation of ISO 9001 so and in ISO 14001 prior to implementing it, an initial review or gap analysis of the organization's processes and products is recommended, to assist in identifying all elements of the current operation and, if possible, future operations, which may interact with the environment, termed "environmental aspects." Environmental aspects can include both direct, such as those used during manufacturing, and indirect, such as raw materials. This review assists the organization in establishing their environmental objectives, goals, and targets, which should ideally be measurable; helps with the development of control and management procedures and processes; and serves to highlight any relevant legal requirements, which can then be built into the policy.

2. *Do* – implement the processes – See clauses 7–9. During this stage, the organization identifies the resources required and works out those members of the organization responsible for implementation and control of the EMS. This includes establishing procedures and processes, although only one documented procedure is specified related to operational control. Other procedures may be required to verify or to reinforce better management control over elements such as documentation control, emergency preparedness and response, and the education of employees, to ensure that they can competently implement the necessary processes and record results. Communication and participation across all levels of the organization, especially top management, is a vital part of the implementation phase, with the effectiveness of the EMS being dependent on active involvement from all employees.

3. *Study (Check)* – measure and monitor the processes and report results – See clause 10. The original word that was used by Shewhart was "check" to indicate a passive activity – meaning strictly a monitoring task. However, later W.E. Deming changed the word to "study" to indicate a proactive activity. So, now when we practice the "Study" phase of the model, we do monitoring the performance, but we also measure (frequently) to ensure (NOT CONTROL) that the organization's environmental targets and objectives are being met. In addition, internal audits are conducted at planned intervals to ascertain whether the EMS meets the user's expectations and whether the processes and procedures are being adequately maintained and monitored.

4. *Act* – take action to improve the performance of EMS based on results – See clauses 4.1 and 4.2. After the study phase, a management review is conducted to ensure that the objectives of the EMS are being met, to what degree and how the results are diffused into the organization for effectiveness. Here, the communication is very important because it serves as the gateway of *doing something* – "taking action" based on the results and evaluating the outcomes accordingly. Remember that an effective evaluation has two phases. The first is to identify and act upon the TGW (Things Gone Wrong),

and the second is to review the TGR (Things Gone Right) so that they may be repeated in the future. Also, the review *must* cover any new regulation and/or other changing circumstances that may need updating in the spirit of continuous improvement. When the review and improvements have been completed, the cycle begins again to improve even further.

When an organization is undertaking the EMS system as a method to improvement, it must recognize that it is quite different from a typical Continual Improvement Process (CIP). CIP in ISO 14001 has three dimensions:

1. *Expansion*: More and more customers demand that an EMS system be implemented in their supplier base. In fact, many have it as a specific requirement of compliance before they do any business with an organization.
2. *Enrichment*: More and more activities, products, processes, emissions, resources, *etc.* are managed by the implemented EMS.
3. *Upgrading*: An improvement of the structural and organizational framework of the EMS, as well as an accumulation of know-how in dealing with business-environmental issues.

Overall, the CIP concept expects the organization to gradually move away from merely operational environmental measures towards a strategic approach on how to deal with environmental challenges.

To be sure, we all have an impact on the environment by the mere fact of existence from day to day. An EMS, in its simplest form, asks us to control our activities so that any environmental impacts are minimized and are part of the sustainability attitude as far as the social, environment, and economic factors are concerned. However, unstructured approach may lead us to improve in the wrong direction or, indeed, may leave us without any clear direction at all. It is tempting to control and minimize those impacts we feel we can tackle easily. Our attitude towards environmental issues is influenced by a topical environmental event, and therefore, we can be influenced to act without thoroughly understanding some of the more complex issues.

We may focus on, and minimize, environmental impacts which are trivial in nature compared with other impacts which are far more significant and require more considered thought processes. Unless a structured approach is taken the organization may focus on what it believes to be its environmental impacts, a belief based upon "feelings" and ease of implementation, rather than empirical data through scientific research. In reality, this does not address real issues but promotes a "green" feel-good factor or perceived enhancement of image – both internal and external to the organization – which is not justified. To be effective, then an organization must move away from these "feelings" approach to a structured system that demands – as a minimum from the organization – an understanding of the concepts behind and strong linkages between

- Identifying all environmental aspects of the organization's activities.
- Using a logical, objective (rather than subjective) methodology to rank such aspects into the order of significant impact upon the environment.

- Focusing the management system to seek to improve upon and minimize such significant environmental impacts.

Benefits of ISO 14001

ISO 14001 was developed primarily to assist companies with a framework for better management control that can result in reducing their environmental impacts. In addition to improvements in performance, organizations can reap a number of economic benefits including higher conformance with legislative and regulatory requirements by adopting the ISO standard. By minimizing the risk of regulatory and environmental liability fines and improving an organization's efficiency, benefits can include a reduction in waste, consumption of resources, and operating costs. Second, as an internationally recognized standard, businesses operating in multiple locations across the globe can leverage their conformance to ISO 14001, eliminating the need for multiple registrations or certifications. Third, there has been a push in the last decade by consumers for companies to adopt better internal controls, making the incorporation of ISO 14001 a smart approach for the long-term viability of businesses. This can provide them with a competitive advantage against companies that do not adopt the standard. This, in turn, can have a positive impact on a company's asset value. It can lead to improved public perceptions of the business, placing them in a better position to operate in the international marketplace. The use of ISO 14001 can demonstrate an innovative and forward-thinking approach to customers and prospective employees. It can increase a business's access to new customers and business partners. In some markets, it can potentially reduce public liability insurance costs. It can serve to reduce trade barriers between registered businesses. There is a growing interest in including certification to ISO 14001 in tenders for public–private partnerships for infrastructure renewal.

Conformity Assessment

ISO 14001 can be used in whole or in part to help an organization (for-profit or not-for-profit) better manage its relationship with the environment. If all the elements of ISO 14001 are incorporated into the management process, the organization may opt to prove that it has achieved full alignment or conformity with the international standard, ISO 14001, by using one of the four recognized options:

1. Make a self-determination and self-declaration.
2. Seek confirmation of its conformance by parties having an interest in the organization, such as customers.
3. Seek confirmation of its self-declaration by a party external to the organization.
4. Seek certification/registration of its EMS by an external organization.

ISO does not control conformity assessment; its mandate is to develop and maintain standards. ISO has a neutral policy on conformity assessment. One option is not better than the next. Each option serves different market needs. The adopting organization decides which option is best for them, in conjunction with their market needs.

- Option 1 is sometimes incorrectly referred to as "self-certify" or "self-certification." This is not an acceptable reference under ISO terms and definitions, for it can lead to confusion in the market. The user is responsible for making their own determination.
- Option 2 is often referred to as a customer or second-party audit, which is an acceptable market term.
- Option 3 is an independent third-party process by an organization that is based on engagement activity and delivered by specially trained practitioners.
- Option 4 is certification, which is another independent third-party process and which has been widely implemented by all types of organizations. Certification is also known in some countries as registration. Service providers of certification or registration are accredited by national accreditation services such as NABCB in India or UKAS in the UK or RAB in the USA.

General Structure of ISO 14001:2015

ISO 14001 was revised in 2015 to bring it up to date with the needs of modern businesses and the latest environmental thinking. It's based on Annex SL, the new high-level structure (HLS) which is a common framework for all ISO management systems. This helps keep consistency, align different management system standards, offer matching sub-clauses against the top-level structure, and apply common language across all standards. It makes it easier for organizations to incorporate their EMS, into core business processes, make efficiencies, and get more involvement from senior management. Perhaps, the biggest difference between the old, and the new standard is the structure. This is because the new edition uses the new Annex SL template. According to ISO, all future management system standards (MSSs) will use this new layout and share the same basic requirements. As a result, all new MSSs will have the same look and feel. A common structure is possible because basic concepts such as management, requirements, policy, planning, performance, objective, process, control, monitoring, measurement, auditing, decision-making, corrective action, and nonconformity are common to all management system standards. A common structure should make it easier for organizations to implement multiple standards because they will all share the same basic language and the same basic requirements.

There have been changes in clause structure and some of the terminology in the new ISO 14001:2015 standard to be in alignment with other standards such as ISO 9001:2015 and ISO 27001:2013. This, however, does not mean that the organization has to change its organization's EMS documentation. There is no requirement to replace the terms used by an organization with the terms used in the new international Standard. Organizations can choose to use terms that suit their business, e.g., "records," "documentation," or "protocols," rather than "documented information."

Layout of ISO 14001:2015

1. Scope
2. Normative References

3. Terms and definitions
4. Context of the organization
    4.1 Understanding the organization and its context
    4.2 Understanding the needs and expectations of interested parties
    4.3 Determining the scope of the environmental management system
    4.4 Environmental management system
5. Leadership
    5.1 Leadership and commitment
    5.2 Environmental policy
    5.3 Organizational roles, responsibilities and authorities
6. Planning
    6.1 Actions to address risks and opportunities
    6.2 Environmental objectives and planning to achieve them
7. Support
    7.1 Resources
    7.2 Competence
    7.3 Awareness
    7.4 Communication
    7.5 Documented information
8. Operation
    8.1 Operational planning and control
    8.2 Emergency preparedness and response
9. Performance evaluation
    9.1 Monitoring, measurement, analysis, and evaluation
    9.2 Internal audit
    9.3 Management review
10. Improvement
    10.1 General
    10.2 Nonconformity and corrective action
    10.3 Continual improvement

1. *Scope*: This section explains the scope of the standard – i.e., what it is for and what it encompasses. It introduces the requirements of an EMS which supports the fundamental "environmental pillar" of sustainability, together with the key intended outcomes of a management system including

    a. Enhancement of performance
    b. Conforming to compliance obligations
    c. Fulfillment of objectives.

    This section also makes clear that any organization claiming compliance with the revised standard should have incorporated all requirements of the standard within their EMS.
2. *Normative references*: As with ISO 14001:2004, there are no normative references associated with ISO 14001:2015. The clause is included simply in order to maintain consistent alignment with the ISO HLS.

3. *Terms and definitions*: This clause lists the terms and definitions that apply to the standard – these are referenced, where necessary, back to other ISO 14001 standards (e.g., ISO 14031:2013). The ISO 14001:2015 standard extends the list of terms and definitions from the ISO 14001:2004 standard, combining the mandated HLS terms and definitions together with the more specific terms and definitions associated with EMSs.

Some of the core concepts of ISO 14001:2015 are shown in Table 1.2.

Clarification of Concepts

The use of the word "any" implies selection or choice. We cannot interchange the words "appropriate" and "applicable" as they are not interchangeable. "Appropriate" means suitable for use and implies some degree of freedom, while "applicable" means relevant or possible to apply and implies that if it can be done, it needs to be done. The word "consider" means it is necessary to think about the topic but it can be excluded,

---

TABLE 1.2

Core Concepts of ISO 14001:2015

Concept	Comment
Context of the organization	The range of issues that can affect, positively or negatively, the way an organization manages its environmental responsibilities
Issues	Issues can be internal or external, and positive or negative and include environmental conditions that either effect or are affected by the organization
Interested parties	More details about considering their needs and expectations, and then deciding whether to adopt any of them as compliance obligations
Leadership	Requirements are specific to top management who is defined as a person or group of people who directs and controls an organization at the highest level
Risk and opportunities	Refined planning process replaces preventive action. Aspects and impacts now part of the risk model
Communication	There are explicit and more detailed requirements for both internal and external communications
Documented information	Replaces documents and records
Operational planning and control	Generally, more detailed requirements, including consideration of procurement, design, and the communication of environmental requirements "consistent with a life cycle perspective"
Performance evaluation	Covers the measurement of EMS, operations that can have a significant environmental impact, operational controls, compliance obligations, and progress towards objectives
Nonconformity and corrective action	The more detailed evaluation of both the nonconformities themselves and corrective actions required

---

whereas "take into account" means it is necessary to think about the topic but it cannot be excluded. "Continual" indicates duration that occurs over a period of time, but with intervals of interruption (unlike "continuous" which indicates duration without interruption). "Continual" is, therefore, the appropriate word to use when referring to improvement. The word "effect" is used to describe the result of a change to the organization. The phrase "environmental impact" refers specifically to the result of a change to the environment. The word "ensure" means the responsibility can be delegated, but not accountability. The terms "interested party" and "stakeholder" are synonyms as they represent the same concept.

Some new terminology used in ISO 14001:2015 and not available in ISO 14001:2004 are as follows:

- The phrase "compliance obligations" replaces the phrase "legal requirements and other requirements as to which the organization subscribes" used in ISO 14001:2004. The intent of this new phrase does not differ from that of the previous edition.
- "Documented information" replaces the nouns "documentation," "documents," and "records" used in ISO 14001:2004. To distinguish the intent of the generic term "documented information," ISO 14001:2015 now uses the phrase "retain documented information as evidence of..." to mean records, and "maintain documented information" to mean documentation other than records. The phrase "as evidence of..." is not a requirement to meet legal evidentiary requirements; its intent is only to indicate objective evidence needs to be retained.
- The phrase "external provider" means an external supplier organization (including a contractor) that provides a product or a service. The change from "identify" to "determine" is done to harmonize with the standardized management system terminology. The word "determine" implies a discovery process that results in knowledge. The intent does not differ from that of previous editions.
- The phrase "intended outcome" is what the organization intends to achieve by implementing its EMS. This includes the enhancement of environmental performance, the fulfillment of compliance obligations, and achievement of environmental objectives. Organizations can set additional intended outcomes for their EMS. For example, consistent with its commitment to the protection of the environment, an organization may establish an intended outcome to work towards sustainable development.
- The phrase "the person doing work under its control" includes persons working for the organization and those working on its behalf for which the organization has the responsibility (e.g., contractors). It replaces the phrase "persons working for it or on its behalf" and "persons working for or on behalf of the organization" used in ISO 14001:2004. The intent of this new phrase does not differ from that of the previous edition. On the other hand, the concept of "Target" used in ISO 14001:2004 is now termed "environment objective."

Clause 4: Context of the Organization

This is a new clause that establishes the context of the EMS and how the business strategy supports this. "Context of the organization" is the clause that underpins the rest of the standard. It gives an organization the opportunity to identify and understand the factors and parties that can affect, either positively or negatively, the EMS. Unlike the old standard, the new one expects you to understand your organization's external context and its internal context before you establish its EMS. This means that you need to identify and understand the external issues and the external environmental conditions that could influence your organization's EMS and the results that it intends to achieve. It also means that you need to identify and understand the internal issues and internal environmental conditions that could influence your EMS and the results it intends to achieve. The new ISO 14001:2015 standard also expects you to identify the interested parties that are relevant to your EMS and to identify their needs and expectations. Once you've done this, it expects you to study these needs and expectations and to figure out which ones have become compliance obligations. But why is all this necessary? It's necessary because your EMS will need to be able to manage all of these influences. Once you understand your context, you're expected to use this knowledge to help you define your EMS and the challenges it must deal with.

4.1 *Understanding the organization and its context*: This clause requires the organization to consider a wide range of potential factors which can impact on the management system, in terms of its structure, scope, implementation, and operation. The areas for consideration include
- Environmental conditions related to climate, air quality, water quality, land use, existing contamination, natural resource availability and biodiversity that can either affect the organization's purpose or be affected by its environmental aspects.
- The external cultural, social, political, legal, regulatory, financial, technological, economic, natural, and competitive circumstances, whether international, national, regional, or local.
- The internal characteristics or conditions of the organization, such as its activities, products and services, strategic direction, culture, and capabilities (i.e., people, knowledge, processes, systems).

4.2 *Understanding the needs and expectations of interested parties*: Clause 4.2 requires the organization to determine the need and expectations of "interested parties," both internal and external. Interested parties could include
- Employees
- Contractors
- Clients/customers
- Suppliers
- Regulators
- Shareholders
- Neighbors

- Non-governmental organizations (NGOs)
- Parent organizations.

What is clear is that while the consideration of context and interested parties need to be relevant to the scope and the standard, the assessment needs to be appropriate and proportionate. What is also clear is that the output from clauses 4.1 and 4.2 is a key input to the assessment and determination of risks and opportunities required in clause 6. There are various methods and approaches which can be used to capture these inputs. As with any significant revision to standards, hopefully, there will be the development of a range of methods and examples for this. Some current examples include

- Internal and external issues as well as key economic and market development issues that may impact an organization. Obviously, a specific organization is probably acutely aware of what is happening in its markets, but this may be undertaken in a very ad-hoc way.
- Technological innovations and developments are also an area critical to your business success and are also probably being monitored and discussed at numerous levels.
- *Regulatory developments* – there is a whole range of external regulations being monitored by your organization; if you miss them, then it could seriously damage your business, or if you capture early intelligence on them, you can more effectively manage any risks.
- *Political and other instabilities* – if, for example, you rely on raw materials from one particular country which experiences major instability, your whole business could be jeopardized; or if there are major environmental concerns regarding a source of materials or goods, this could have significant reputational consequences.
- *Organizational culture and attitudes* – an effective and motivated workforce will give you positive impacts, and many organizations canvas feedback from employees.
- *Internal and external parties* – stakeholder engagement exercises are already widely used to consult with interested parties and map out concerns and issues, more often used by larger organizations engaging with corporate social responsibility initiatives,
  - Consultation meetings with neighborhoods and NGOs (non-government organizations) on environment, planning, and development issues are often used by major industrial plants with significant HSE (Health, Safety, and Environment) risks.
  - Meetings and other interactions with regulators can encompass, for example, critical issues on product specifications and conformity from an environmental perspective, as well as issues with regulators on compliance and developing compliance against emerging requirements and standards.
  - Employee meetings, consultations, and feedback activities – this should be happening already, but maybe this will prompt more

efforts to improve an area which has been at risk of "lip service" to ISO 14001:2004.
– Supplier reviews and relationship management – many organizations are trying to get much more mutual benefit from the supplier–client relationships which are critical to mutual success.
- Client/customer reviews and relationship management – of course, this is a fundamental pillar of all standards and a key to success.
- It may be that when you reflect on how you capture key issues, and how many interested parties you engage with already, you may be pleasantly surprised. It may be that you only engage with a limited number of internal and external parties, but now is the time to start thinking about whether that is enough, and whether you are missing some good opportunities. There will be many ways in which to capture this – and hopefully, some improved and new approaches might emerge as this part of the standard is considered. Approaches could include
  – Summary information from the range of existing approaches used as listed above (e.g., a brief report).
  – The information summarized as part of inputs to risk and opportunity registers (e.g., for ISO 14001, this could be an additional process in the identification of environmental aspects and impacts).
  – Recorded in a simple spreadsheet.
  – Logged and maintained in a database.
  – Captured and recorded through key meetings.
  – These clauses are asking organizations to think clearly and logically about what can internally and externally affect their management systems, and to be in a position to show that this information is being monitored and reviewed. It also requires organizations to elevate the discussions to the highest levels, since capturing the above range of information is hard to achieve without a high-level approach.

4.3 *Determining the scope of the EMS*: This clause should be familiar to most organizations, since ISO 14001:2004 clause 4.1 clearly requires the definition of the scope of the management system. For ISO 14001:2015, the scoping requirements have become clearer and stronger, and require the organization to consider the inputs from clauses 4.1 and 4.2, along with the products and services being delivered. This should encourage a clearer and more logical approach to scope, driven by external and internal requirements – it should not be used to exclude activities, processes, or locations which have significant environmental aspects and impacts and should not be used to avoid areas with clear compliance obligations. The scope should be clearly documented and made publicly available. These clearer requirements on scoping will drive clarity in the thinking of organizations in scoping the management system. Certification bodies will, as before, look at how an organization has defined its scope, ensuring that this is both appropriate and is reflected accurately by the management system and also in the scope of any certificate issued.

4.4 *Environmental management system:* This clause basically states that the organization needs to establish, implement, maintain, and continually improve a management system in order to achieve its intended outcomes, including enhancement of environmental performance. This should also be familiar to organizations which implement management systems in order to deliver compliance and improvement. This clause is also more focused in requiring organizations to understand more about the range of processes relevant to the scope of the management system. The term "process" is defined as "a set of interrelated or interacting activities which transforms inputs into outputs."

For those who are committed to a management system which is at the core of your business, this will probably be an integral part of your management system. You may, however, need to review how effectively you connect those processes and understand the influence and impact of those processes on each other and on the business. This should also elevate the system in terms of its importance and value to the business because it should drive more meaningful analysis of the key business processes and critical aspects of those processes. In practical terms, it will require an organization to more fully analyze its processes and ensure that there is a good understanding of how they interact with each other – and not operate as isolated procedures without overlap.

Clause 4 introduces some significant innovations to the management system world and could represent a challenge to some organizations who have not viewed the management system as essential to the business, focused as it is on raising management systems to a higher level and to be more central to the way an organization works – an approach which is entirely correct and logical.

Clause 5: Leadership

This clause is all about the role of "top management" which is the person or group of people who directs and controls the organization at the highest level. The purpose is to demonstrate leadership and commitment by integrating environmental management into business processes. Top management must demonstrate a greater involvement in the management system and need to establish the environmental policy, which can include commitments specific to an organization's context beyond those directly required, such as the "protection of the environment." There is a greater focus on top management to commit to continual improvement of the EMS. Communication is key, and top management has a responsibility to ensure the EMS is made available, communicated, maintained, and understood by all parties. Finally, top management needs to assign relevant responsibilities and authorities, highlighting two particular roles concerning EMS conformance to ISO 14001 and reporting on EMS performance.

5.1 *Leadership and commitment:* This clause encompasses a range of key activities which top management needs in order to "demonstrate leadership and commitment with respect to the management system." Therein lies one of the innovations delivered by the common HLS – top management must show leadership of the management system rather than just demonstrate the commitment to it. The standard is driving the oversight of the management system to the highest level of management and making it a key component of the organization and its core business processes and activities. It doesn't mean that the leadership has to be able to regurgitate the policy or recite the objectives and targets – what it means is that an internal or external interested party should feel entitled to have a discussion with leadership about core and critical aspects of the business, because these are at the heart of the management system. This sub-clause is a significant innovation to the structure of management systems but should be viewed as a "positive challenge" to organizations and an opportunity to enhance the role of the EMS and place it at the center of the business.

5.2 *Environmental policy:* The environmental policy is an important document because it acts as the driver for the organization. It provides the direction and formally establishes goals and commitment. Top management should ensure that the policy is appropriate, compatible with the strategic direction and not a bland statement that could apply to any business. It should provide clear direction to allow meaningful objectives to be set that align with it. The new standard focuses on the commitment to "protection of the environment" rather than solely addressing "prevention of pollution" in the 2004 edition. This indicates a broader environmental view and is more in line with current and future environmental challenges. Commitments to protect the environment can, in addition to prevention of pollution, also include climate change mitigation and adaptation, sustainable resource use, and protection of biodiversity and ecosystems. The policy needs to be communicated to all employees, and they need to understand the part they have in its deployment. The policy must be documented and available externally.

5.3 *Organizational roles, responsibilities, and authorities*: For a system to function effectively, those involved need to be fully aware of what their role is. Top management must ensure that key responsibilities and authorities are clearly defined and that everybody involved understands their roles. Defining roles is a function of planning, ensuring awareness can then be achieved through communication and training. It is common for organizations to use job descriptions or procedures to define responsibilities and authorities. In ISO 14001:2015, top management is more directly identified as being responsible for ensuring that these aspects of the system are properly assigned, communicated, and understood. The specific role of a management representative has been removed. However, the standard still contains all of the key activities and responsibilities of that previously identified role, but these now lie more directly within the core structure of the organization – including top management. This has a positive implication for the EMS because there is a clear expectation for consistent and appropriate ownership from top to bottom within an organization.

Clause 5 contains much familiar content, but with greater emphasis on leadership and commitment, and the expectation that top management will be more actively engaged with the management system.

Clause 6: Planning

Unlike the old standard, the new ISO 14001 standard expects you to determine risks and opportunities. So, what does this mean, and what does the new standard expect you to do? It expects you to start by establishing a risk planning process. It then expects you to use this process to identify risks and opportunities related to your organization's unique context, its interested parties, its compliance obligations, and its environmental aspects. It then expects you to define actions to address all of these risks and opportunities. And to make sure that these actions will actually be carried out, it asks you to make these actions an integral part of your EMS, and then to implement, control, evaluate, and review the effectiveness of these actions and these processes. While risk planning is now an integral part of the new ISO 14001 standard, it does not actually expect you to implement a formal risk management process.

6.1 *Actions to address risks and opportunities*: In basic terms, this clause requires the organization to
- Consider in planning the EMS the context of the organization and the scope of the system 6.1.1.
- Determine risk and opportunities relating to environmental aspects 6.1.2, compliance obligations 6.1.3, and other issues and requirements identified in 4.1 and 4.2 (6.1.1).
- Also, consider potential emergency situations which could arise and constitute risk (6.1.1).
- In addition, and as required already by ISO 14001:2004, determine the range of environmental aspects and impacts, and determine those impacts which are of significance to the organization within the defined scope (6.1.2).
- Consider all compliance obligations applicable to the organization and how these may present threats or opportunities (6.1.3).
- The organization then needs to consider appropriate actions to address the significant aspects/impacts (6.1.2), compliance obligations (6.1.3), and risks and opportunities identified (6.1.1).

Risks and Opportunities   The ISO 14001 standard also introduces the concept of "considering a life cycle perspective" for its products, services, and activities (see Figure 1.1).

This makes the previous concepts of the upstream and downstream aspects clearer, and also introduces language now in common use across other standards as well as Corporate Social Responsibility (CSR) and product assessment standards. The overall strength of this clause lies in both introducing the principles of risk and opportunity to management systems standards via the HLS, and by connecting it very clearly to the processes defined under clause 4. A well-established approach to managing this range of inputs, risk analysis, and prioritization already implemented

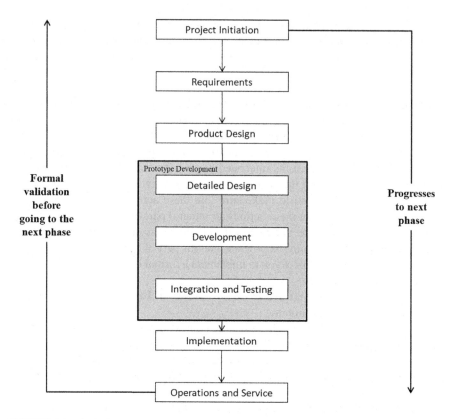

FIGURE 1.1 A depiction of a typical project life cycle.

by many organizations is the use of risk registers, which if properly managed and implemented can effectively identify and assess risks and opportunities across a wide range of areas and issues. There will also be other approaches which result from the various relevant clauses of 14001 (e.g., the results from clauses 4.1 and 4.2 and the requirements of 6.1.1, 6.1.2, 6.1.3 and 6.1.4) along with management of change, with an overall analysis and review resulting in objectives, targets, and plans. The depth and complexity of approach will depend significantly on the size and complexity of the organization, as well as other factors which could include the level of external regulation, existing requirements for public reporting, shareholder interests, public profile, numbers and types of customers, range and types of suppliers. Hence, there will be a range of approaches that will be appropriate for the wide spectrum of organizations.

6.2 *Environmental objectives and planning to achieve them*: This clause requires the organization to establish environmental objectives and plans, ensuring that these are clear, measurable, monitored, communicated, updated, and resourced. As part of the planning process, top management needs to set environmental objectives driven by the outputs from the analysis of risks arising from threats and opportunities (i.e., the range of activities undertaken in 6.1), with the aim of delivering compliance, performance improvement, and effective risk management. Objectives should be consistent with the environmental policy and be capable of being measured. Documented information needs to keep in relation to objectives, and there will need to be evidence regarding monitoring of achievement.

Preventive Action   The new ISO 14001 standard no longer uses the term "preventive action." We're now expected to use risk planning concepts and to think of the entire EMS as a system of preventive action. ISO 14001:2015 Section A.10.1 says there is no longer a single clause on preventive action because "One of the key purposes of an environmental management system is to act as a preventive tool. This concept of preventive action is now captured in 4.1 (i.e., understanding the organization and its context) and 6.1 (i.e., actions to address risks and opportunities)." So, according to the new standard, these two sets of requirements cover the old concept of preventive action. Evidently, once we realize that the entire EMS can be used to manage risks and opportunities, we no longer need a separate clause on preventive action. It's redundant.

Clause 7: Support
This clause is all about the execution of the plans and processes that enable an organization to meet its EMS. Simply expressed, this is a very powerful requirement covering all EMS resource needs. Organizations will need to determine the necessary competence of people doing work that, under its control, affects its environmental performance, its ability to fulfill its compliance obligations, and ensure they receive the appropriate training. In addition, organizations need to ensure that all people doing work under the organization's control are aware of the environmental policy, how their work may impact this, and implications of not conforming with the EMS. Finally, there are the requirements for "documented information" which relate to the creation, updating, and control of specific data.

7.1 *Resources*: The main intention behind this general requirement is that the organization must determine and provide the resources needed for the establishment, implementation, maintenance and continual improvement of the environmental management. That is, covering all aspects of people and infrastructure. Here, we must note that while *organizational knowledge* is not directly mentioned in the ISO 14001 standard, the ISO 9001:2015 clause 7.1.6 standards cover it with three distinct subdivisions which relate to ensuring that the organization understands internal and external knowledge needs and can demonstrate how this is managed. This could also include knowledge management of resources, and ensuring that there is effective succession planning for personnel, and processes for capturing individual and group knowledge. It isn't a documented requirement of ISO 14001, but it is relevant and useful as a general principle.

7.2 *Competence*: In order to determine competence, competence criteria need to be established for each function and role relevant to the EMS. This can then be used to assess existing competence and determine future needs. Where criteria are not met, some action is required to fill the gap. Training or reassignment may even be necessary. Retained documented information is required to be able to demonstrate competence. Recruitment and induction programs, training plans, skills tests, and staff appraisals often provide evidence of competence and their assessment. Competency requirements are often included in recruitment notices and job descriptions. The standard is clear that documented information is required as evidence of competence.

7.3 *Awareness*: Personnel needs to be made aware of the environmental policy, significant aspects and impacts of relevance to their activities, how they contribute to the environmental objectives, environmental performance and compliance obligations, and the implications of failures in compliance.

7.4 *Communication*: Effective communication is essential for a management system. Top management needs to ensure that mechanisms are in place to facilitate this. It should be recognized that communication is two way and will not only need to cover what is required but also what was achieved. With ISO 14001:2015, the importance of internal communications and external communications is emphasized. This is a natural legacy of the existing ISO 14001:2004 and the importance of interested parties in environmental issues. This sub-clause also makes very clear the importance of ensuring in relation to environmental reporting and associated communications that the organization shall "ensure that environmental information communicated is consistent with information generated within the environmental management system, and is reliable." This is an excellent addition and consistent with other corporate reporting standards. It also emphasizes the need to plan and implement a process for communications along with the familiar "who, what, when, how" principles.

7.5 *Documented information*: The new ISO 14001:2015 standard has also eliminated the long-standing distinction between documents and records. Now they're both referred to as "documented information." Why ISO chose to abandon two common sense concepts and replace them with one that is needlessly awkward and esoteric is not entirely clear. According to ISO's definition, the term "documented information" refers to information that must be controlled and maintained. So, whenever ISO 14001:2015 uses the term "documented information" it implicitly expects you to control and maintain that information and its supporting medium. However, this isn't the whole story. The annex to the new ISO 14001:2015 standard (A.3) further says that "this international standard now uses the phrase 'retain documented information as evidence of' to mean records, and 'maintain documented information' to mean documentation other than records." So, whenever the new ISO 14001 standard refers to documented information and it asks you to maintain this information, it is talking about what used to be referred to as documents, and whenever it asks you to retain this information, it is talking about what used to be called records. So sometimes documented information must be maintained, and sometimes, it must be retained (contrary to what ISO's official definition says). So, while the official definition of the term "documented information" abandons the distinction between documents and records, through the use of the words "maintain" and "retain" and because of what this means (according to Annex A), the main body of the standard actually restores this distinction. In other words, while documents and records were officially kicked out the front door, they were actually allowed back in through the back door.

Procedures   The old ISO 14001 standard asked organizations to establish a wide range of procedures. These included an environmental aspects procedure, a legal requirements management procedure, an awareness procedure, a communications procedure, a documents procedure, an operational procedure, an emergency preparedness and response procedure, a monitoring and measurement procedure, a compliance evaluation procedure, a nonconformity management procedure, a record-keeping procedure, and an audit procedure. Now, only one procedure is left. The new ISO 14001:2015 standard asks you to establish an emergency preparedness and response procedure in Section 8.2 and that's the only one. Instead of asking you to write procedures, the new standard expects you to maintain and control a wide range of documents (i.e., documented information). Since the new standard doesn't tell you what to call these documents, you can call them procedures if you like. And, of course, you still need to have documents except that now they're called "documented information." So, while on the surface, this looks like a radical change, in reality, it is not so. Mandatory documents and records required by ISO 14001:2015 are the documents you need to produce if you want to be compliant with ISO 14001:2015. Here, we provide the applicable clauses for the mandatory documents:

- Scope of the EMS (clause 4.3)
- Environmental policy (clause 5.2)
- Risk and opportunities to be addressed and processes needed (clause 6.1.1)
- Criteria for the evaluation of significant environmental aspects (clause 6.1.2)
- Environmental aspects with associated environmental impacts (clause 6.1.2)
- Significant environmental aspects (clause 6.1.2)
- Environmental objectives and plans for achieving them (clause 6.2)
- Operational control (clause 8.1)
- Emergency preparedness and response (clause 8.2)

And, here are the mandatory records:

- Compliance obligations record (clause 6.1.3)
- Records of training, skills, experience, and qualifications (clause 7.2)
- Evidence of communication (clause 7.4)
- Monitoring and measurement results (clause 9.1.1)
- Internal audit program (clause 9.2)
- Results of internal audits (clause 9.2)
- Results of the management review (clause 9.3)
- Results of corrective actions (clause 10.1).

Non-mandatory Documents

There are numerous non-mandatory documents that can be used for ISO 14001 implementation. However, the following non-mandatory documents are the most utilized:

- Procedure for determining the context of the organization and interested parties (clauses 4.1 and 4.2)
- Procedure for identification and evaluation of environmental aspects and risks (clauses 6.1.1 and 6.1.2)
- Competence, training, and awareness procedure (clauses 7.2 and 7.3)
- Procedure for communication (clause 7.4)
- Procedure for document and record control (clause 7.5)
- Procedure for internal audit (clause 9.2)
- Procedure for management review (clause 9.3)
- Procedure for management of nonconformities and corrective actions (clause 10.2).

Clause 8: Operation

This clause deals with the execution of the plans and processes that enable the organization to meet its environmental objectives. There are specific requirements that relate to the control or influence exercised over outsourced processes and the requirement to consider certain operational aspects "consistent with a life cycle perspective." This means giving serious consideration to how actual or potential environmental impacts happening upstream and downstream of an organization's site-based operations are influenced or (where possible) controlled. Finally, the clause also covers the procurement of products and services, as well as controls to ensure that environmental requirements relating to design, delivery, use and end-of-life treatment of an organization's products and services are considered at an appropriate stage.

8.1 *Operational planning and control*: The overall purpose of operational planning and control is to ensure that processes are in place to meet the EMS requirements and to implement actions identified in 6.1 and 6.2. There are some clearer and stronger requirements relating to outsourced processes and control of changes. In addition, requirements around the life cycle perspective approach are defined in more detail, covering the key elements of
- Environmental requirements for procurement of products and services
- Establishing controls to ensure environmental requirements are addressed in the design and development phase
- Communicating environmental requirements to providers (including suppliers, contractors, and others)
- Providing key environmental information on products and services in the context of the life cycle (e.g., end-of-life information).

The organization needs to determine and evaluate the level of control and influence over the different life cycle elements, based on the context of the organization and the consideration of significant environmental aspects, compliance obligations, and risks associated with threats and opportunities.

Overall, ISO 14001:2015 requires a structured approach to all aspects of the products and services with a strong reference point to life cycle perspective. As discussed under clause 7.5, there is no specific requirement for documented procedures in ISO 14001:2015, but there is a clear requirement for ensuring that there is documented information to provide assurance that the processes are in place and implemented effectively. That requirement could cover process maps, procedures, specifications, forms, records, data and other information across any media.

8.2 *Emergency preparedness and response:* This clause is clear in requiring the organization to establish, implement, and maintain processes needed to handle potential emergency situations identified in 6.1.1. The more detailed requirements cover the need to ensure

- That the organization plans actions to mitigate or prevent environmental consequences.
- That the organization responds to actual emergency situations.
- That the organization takes appropriate and applicable action to prevent or mitigate the consequences of emergency situation.
- That the organization conducts periodic testing of any procedures, plans, and response mechanisms.
- That the organization conducts periodic reviews and updates of procedures and plans based on experience.
- That the organization has a provision of communicating relevant information and training to the applicable stakeholders.

Clause 9: Performance evaluation

This is all about measuring and evaluating your EMS to ensure that it is effective and it helps you to continually improve. You will need to consider what should be measured, the methods employed, and when data should be analyzed and reported on. As a general recommendation, organizations should determine what information they need to evaluate environmental performance and effectiveness. Internal audits will need to be carried out, and there are certain "audit criteria" that are defined to ensure that the results of these audits are reported to relevant management. Finally, management reviews will need to be carried out, and "documented information" must be kept as evidence.

9.1 *Monitoring, measurement, analysis, and evaluation:* This sub-clause encompasses two key areas:
  - Monitoring, measurement, analysis, and evaluation of environmental performance and the effectiveness of the system.
  - Evaluation of compliance with all legal and other obligations.
  The range of monitoring and measurement required needs to be determined for those processes and activities which relate to significant environmental aspects/impacts, environmental objectives, key areas of operational control and processes, and also for evaluating the meeting of compliance

obligations. For the monitoring and measurement determined as required, the organization also needs to determine key criteria and requirements, including

- Methods for monitoring, measurement, analysis, and evaluation.
- Key performance indicators and performance evaluation metrics; when, where, how, and by whom the monitoring, measurement, evaluation, and analysis is carried out.
- Specification, management, and maintenance of key monitoring equipment and data handling processes.

The output from these activities provides key inputs for a range of other elements of the EMS, including management review, and in determining the internal and external communications required on the EMS and its performance.

The other key aspect of this sub-clause is that the organization will need to demonstrate how it evaluates compliance with other requirements. Most organizations fulfill this clause via their internal audit processes, but other compliance audits, checks, and reviews can be used. The organization should define its processes for evaluating compliance with legal and other requirements and must maintain documented information relating to these activities. The process must cover

- Frequency of evaluation
- Evaluation approach
- Maintain knowledge on compliance status.

This area is similar to the requirements under ISO 14001:2004, but with clearer and more detailed requirements. As with ISO 14001:2004, this is not about reviewing which compliance obligations are applicable to the organization, it is about evaluating actual compliance with the range of compliance obligations applicable to the organization.

9.2 *Internal audit*: Internal audits have always been a key element of ISO 14001 in helping to assess the effectiveness of the EMS. An audit program needs to be established to ensure that all processes are audited at the required frequency, the focus being on those most critical to the business. To ensure that internal audits are consistent and thorough, a clear objective and scope should be defined for each audit. This will also assist with auditor selection to ensure objectivity and impartiality. To get the best results, auditors should have a working knowledge of what is to be audited, but management must act on audit results. This is often limited to corrective action relating to any nonconformities that are found, but there also needs to be consideration of underlying causes and more extensive actions to mitigate or eliminate risk. Follow-up activities should be performed to ensure that the action taken as a result of an audit is effective. This clause is largely the same as ISO 14001:2004.

9.3 *Management review*: The main aim of management review is to ensure the continuing suitability, adequacy, and effectiveness of the quality management system. Only through conducting the review at sufficient intervals (remember management review does not have to be just one meeting, held once per year), providing adequate information and ensuring the right people are involved can this aim be achieved. The standard details the minimum inputs to the review process. Top management should also use the review as an opportunity to identify improvements that can be made and/or any changes required, including the resources needed. The input to management review should include information on

- Status of previous actions from management reviews
- Changes in internal/external inputs, significant aspects/impacts, and compliance obligations
- Achievement and progress on environmental objectives
- Information on environmental performance
- Communications from external interested parties
- Opportunities for continual improvement
- Adequacy of resources for the EMS

The output from the management review should include any decisions and actions related to

- Conclusions on the suitability, adequacy, and effectiveness of the system
- Continual improvement opportunities
- Changes to the EMS, including resources
- Actions relating to objectives not achieved
- Implications for the strategic direction of the organization
- Documented information pertaining to the management review is required to be retained.

This clause is largely the same as ISO 14001:2004, but with some broader topics and alignment with the new language of risks and opportunities, and the context of the organization.

Clause 10: Improvement

This clause requires organizations to determine and identify opportunities for continual improvement of the EMS. The requirement for continual improvement has been extended to ensure that the suitability and adequacy of the EMS – as well as its effectiveness – are considered in the light of enhanced environmental performance. There are some actions that are required and cover the handling of corrective actions. First, organizations need to react to the nonconformities and take action. Second, they need to identify whether similar nonconformities exist or could potentially occur. This clause requires organizations to determine and identify opportunities for continual improvement of the EMS. There is a requirement to actively look out for opportunities to improve processes, products, or services, particularly with future customer requirements in mind.

10.1 *General*: This states that the organization shall determine opportunities for improvement and implement necessary actions to achieve intended outcomes.

10.2 *Nonconformity and corrective action*: The main aim of the corrective action process is to eliminate the causes of actual problems so as to avoid recurrence of those problems. It is a reactive process in that it is triggered after an undesired event (e.g., a pollution event). In essence, the process uses the principles of root cause analysis. A basic approach to problem-solving is "cause" and "effect," and it is the cause that needs to be eliminated. Action taken should be appropriate and proportionate to the impact of the nonconformity. As part of the corrective action process, the effectiveness of action taken must be checked to ensure it is effective. For this clause on nonconformity and corrective action, much of the content is familiar and similar to ISO 14001:2004, but the term "preventive action" has now been completely deleted from the standard. This is because the new HLS is built on the fundamental principles of risk management, which embodies the need to identify risks and manage those risks, with the ultimate goal of risk elimination. The overall approach is one of mitigating and where possible eliminating risk, with the use of corrective action to deal with the impacts of realized risks.

10.3 *Continual improvement*: This sub-clause of ISO 14001:2015 effectively summarizes the key aim of an EMS: to continually improve the suitability, adequacy, and effectiveness of the EMS to enhance environmental performance. This was also embodied in ISO 14001:2004 but is separately stated in ISO 14001:2015. Improvement does not have to take place in all areas of the business at the same time. The focus should be relevant to risks and benefits. Improvement can be incremental (small changes) or breakthrough (new technology). In reality, both methods will be used at some point in time.

Other Clarifications and Modifications   The old ISO 14001 standard asked you to "define and document the scope of its environmental management system" (4.1), but it didn't say anything about how this should be done. The new ISO standard clarifies how this ought to be done (4.3). It now asks you to consider your compliance obligations, your corporate context, your physical boundaries, your products and services, your activities and functions, and your authorities and abilities when you define the scope of your EMS. And it asks you to include all products, services, and activities that have significant environmental aspects. The new term "compliance obligation" has replaced the rather cumbersome phrase, "legal requirements and other requirements to which the organization subscribes." However, the meaning is the same. There are two kinds of compliance obligations: (a) mandatory compliance obligations and (b) voluntary compliance obligations. Mandatory compliance obligations include laws and regulations, while voluntary compliance obligations include contractual commitments, community and industry standards, ethical codes of conduct, and good governance guidelines. A voluntary obligation becomes mandatory once you decide to comply with it. The new standard no longer refers to environmental targets. According to Section A.6.2, "The concept of "target" used in prior editions of this International Standard is captured within the definition of "environmental objective."" You can, of course, still set targets and call them targets if you wish. The only real difference is that the new ISO 14001 standard thinks of a target as a type of objective.

Management of change is an important part of maintaining the EMS that ensures the organization can achieve the intended outcomes of its EMS on an ongoing basis. Management of change is addressed in various requirements in ISO 14001:2015 standard, such as maintaining the environmental management system (4.4), environmental aspects (6.1.2), internal communication (7.4.2), operational control (8.1), internal audit program (9.2.2), and management review (9.3).

As part of managing change, the organization should address planned and unplanned changes to ensure that the unintended consequences of these changes do not have a negative effect on the intended outcomes of the EMS. Examples of change include but not limited to planned changes to products, processes, operations, equipment, or facilities; changes in staff or external providers, including contractors; new information related to environmental aspects, environmental impacts, and related technologies; and changes in compliance obligations.

ISO 14001:2015 has introduced the requirement for a "life cycle perspective" (see Figure 1.1) in EMSs. The new standard does not require a formal life cycle analysis or quantification, but does require organizations to look upstream and downstream of the processes performed on-site and try to reduce environmental impacts. Specifically, the life cycle perspective is related to the organization's environmental aspects and impacts. It requires careful consideration of the life cycle stages that the organization can control or influence, including acquisition of raw materials, production and transportation, use and maintenance, and recycling or disposal. In doing this, the organization needs to create records as evidence that they have considered each life cycle stage.

The standard also requires the organization to provide information to its external service providers and contractors about the potentially significant environmental impacts of its products and services. It must also consider the need to provide this information to transporters, end-users, and disposal facilities. By providing this information, the organization can potentially prevent or mitigate adverse environmental impacts during these life cycle stages.

The life cycle perspective (see Figure 1.1) can be applied in choosing, for example:

- Raw materials (environmental impacts of their production, distance transported, and mode of transport)
- Products to manufacture and offer for sale (same considerations as well as disposal or recycling options at end of life)
- Services used by organization (environmental credentials, chemicals used, waste generated)
- Equipment purchases (distance transported, options for recycling at end of life, waste generated in their use).

Why has this life cycle requirement been added to the standard? The introduction to the new standard explains that a life cycle perspective can be used to benefit the environment in areas where the organization has "control or influence" and also "prevents environmental impacts from being unintentionally shifted elsewhere within the life cycle." Life cycle considerations were largely ignored by the old standard. Now they are central. ISO 14001 now expects you to use a life cycle perspective to "identify the environmental aspects and associated environmental impacts of its activities, products and services that it can control and those that it can influence." The term "management representative" has been officially dropped. The management duties and responsibilities that were previously assigned to someone called a "management representative" may now be assigned either to one person or to many. Of course, you may continue to use this job title if you wish.

Things That an Auditor Should Be Concerned With

Any standard requires some key issues that have to be audited for validation and confirmation as to whether they are being performed and monitor for acceptability and continuous improvement. The ISO 14001 is no different.

In this section of the chapter, we have provided the reader (potential auditee and/ or auditor) with some background of what the standard calls for and the appropriate usage in generating an appropriate and applicable checklist for conformance.

ISO 18001/45001

General Comments

While both standards are targeted towards improvements in working conditions, ISO 45001 takes a proactive approach to risk control that starts with the incorporation of health and safety in the overall management system of the organization, thus driving top management to have a stronger leadership role in the safety and health program.

On the other hand, OHSAS 18001 takes a reactive approach of hazard control by delegation of hazard control responsibilities to safety management personnel rather than integrating the responsibilities into the overall management system of the company.

Understanding the differences between the programs is important for employers as they move into the new system and explore the organizational possibilities. Below are several of the main differences between the two standards.

Management Commitment

In ISO 45001, management commitment is central to the standard's effectiveness and integration. The shift in the new standard is towards managerial ownership. The safety culture of the organization is to be supported by the engagement of management with workers and demonstrated by a top-down emphasis. Instead of providing oversight of the program, management should be true safety leaders. This means an active, participatory role in the organization's safety and health for corporate officers, directors, and those in management. Protection of workers and performance improvements are roles of leadership under the new ISO 45001.

Worker Involvement

Workers also have broader participation in the new standard, with employees working with management to implement the safety management system (SMS). Employees should be provided training and education to identify risks and help the company create a successful safety program. Internal audits and risk assessment results should be openly shared with workers and allow for employee input. According to ISO 45001, the responsibility of safety management belongs to everyone in the organization.

Risk versus Hazard

ISO 45001 follows a preventative process, which requires hazard risks to be evaluated and remedied, as opposed to hazard control, under OHSAS 18001. Think of the new standard as proactive, rather than reactive. In adopting ISO 45001, your organization will find and identify potential hazard risks before they cause accidents and injuries. Audits, job safety analyses, and monitoring of workplace conditions will be vital to ensure the proactive approach prescribed by ISO 45001.

Structure

One obvious and important difference between ISO 45001 and OHSAS 18001 is the structure. The new standard is based on Annex SL, which replaced ISO Guide 83, and applies a universal structure, terminology, and definitions. You're probably well familiar with this structure if you also use other systems such as ISO 9001 and ISO 14001. Through using the same structure, multiple management systems are easier to implement in a more streamlined and efficient way.

Ultimately, ISO 45001 can be best summarized as a whole-company, proactive approach to incorporating a safety culture. It is a framework that can take your organization to the next level in safety and health.

ISO 18001

General Comments

Whereas the ISO 9001 defines quality requirements for all organizations, the ISO 18001 and the newer version of ISO 45001 are specific standards that deal with Occupational Health & Safety Management System (OH&SMS). The beginning of these standards was the OHSAS 18001 – Occupational Health and Safety Assessment Series (officially BS OHSAS 18001). The intent of this British standard was to make sure that there exists a standard for *compliance* that enabled the organization to demonstrate that occupational health and safety was present and followed.

As important the ISO 18001 was and accepted by organizations throughout the world, the British Standard Institute (BSI) canceled BS OHSAS 18001 to adopt ISO 45001 as BS ISO 45001 with some clarifications and additional important items for improvement. ISO 45001 was published in March 2018 by the ISO. Organizations that are certified to BS OHSAS 18001 can migrate to ISO 45001 by March 2021 if they want to retain a recognized certification.

"What is OHSAS 18001?" Is there a simple answer to this question? OHSAS 18001, Occupational Health and Safety Management Systems – Requirements (officially BS OHSAS 18001, but often mistakenly called ISO 18001) is a British Standard for occupational health and safety management systems that is recognized and implemented worldwide. The most recent version of this standard was published in 2007 and is referred to as "OHSAS 18001:2007."

Hopefully, in this section, we will address the basics of OHSAS 18001 and help you demystify and discover what the OHSAS 18001 requirements are, and give you a guide on what needs to be done to implement an occupational health & safety management system and become certified through the appropriate and applicable audit.

Benefits of OHSAS

So, let us start with the simple question of: What is an occupational health and safety management system? An occupational health & safety management system, often called an OH&SMS is comprised of the policies, processes, plans, practices, and records that define the rules governing how your company takes care about occupational health and safety. This system needs to be tailored to your particular company, because only your company will have the exact legal requirements and occupational health and safety hazards that match your specific business processes. However, the OHSAS 18001 requirements provide a framework and guidelines for creating your occupational health & safety management system so that you do not miss important elements needed for an OH&SMS to be successful.

Therefore, taking care of occupational health and safety and preventing injuries in the work place are among the most important challenges facing businesses today. One of the biggest benefits of implementing an OH&SMS is the recognition that comes with being among those businesses that care for its employees' health and safety. This can bring better relationships with customers, the public, and the community at large for your company, but it also brings other benefits.

Along with the good public image, many companies can save money through the implementation of an occupational health & safety management system. This can be achieved through reducing incidents that can result in workers' injuries, and being able to obtain insurance at a more reasonable cost. This improvement in cost control is a benefit that cannot be overlooked when making the decision to implement an occupational health and safety management system.

Again, to appreciate the standard, one must understand the basis for its existence, which fundamentally is the basis for the standard. That is, to evaluate the following three categories for improvement:

1. *Hazard identification*: The process of recognizing that a hazard exists (source or situation with the potential to cause harm in terms of human injury or ill health)
2. *Risk assessment*: The process of evaluating the risk arising from the hazard (combination of the likelihood of a hazardous event or exposure and the severity of injury or ill health that can be caused by the event of exposure)
3. *Determination of applicable controls*: Measures relevant to eliminate or reduce risk to an acceptable level. Measures are based on the hierarchy of control measures.

Requirements

Therefore, in order to facilitate this evaluation, the OHSAS 18001 structure is split into four sections. The first three are introductory, with the last section, split into six sub-sections, containing the requirements for the EMS. Here is what the six sub-sections are about:

Section 4.1: General requirements – This section provides an overall statement that the occupational health & safety management system needs to be established, documented, implemented, maintained, and continually improved according to the requirements of the OHSAS 18001 standard. This highlights that the OH&SMS is not a one-time activity to be done and then forgotten, but instead is intended to be maintained to promote improvement.

Section 4.2: OH&S Policy – The occupational health & safety policy helps to set the overall goals to meet the scope of the occupational health & safety management system. The policy includes the company's commitment to comply with legal requirements, prevent injury and bad health, and continually improve. It also provides the overall framework to set the objectives for the OH&SMS.

Section 4.3: Planning – There are three parts to the planning process for the OHSAS 18001. First, the company needs to identify the hazards and assess the risks for every work place. Next, the company needs to identify the legal and other requirements that pertain to the identified hazards and operational processes, and ensure that they are understood and implemented. Lastly, objectives and programs for improvement of the occupational health and safety management system need to be put in place with appropriate resources to accomplish the goals.

Section 4.4: Implementation and operation – This section has many elements to consider, starting with the assignment of resources, roles, responsibilities, and authorities. Once this is in place you must ensure that competence, training, awareness, and communication (both internal and external to the company) are established for the functioning of the OH&SMS. Documentation and control of documents is required to ensure consistency, as is putting in place operational controls and processes for emergency preparedness and response to ensure that there is uniformity where required.

Section 4.5: Checking – The monitoring and measurement, including evaluation of compliance with legal and other requirements, are necessary to ensure that decisions can be made. Part of this is dealing with nonconformity, corrective action, preventive action, and auditing the processes in place. Without these elements, and the records associated with them, it is almost impossible to tell if things are going according to plan.

Section 4.6: Management review – Hand in hand with the records from the checking requirement is this requirement for management to review the recorded outputs in order to ensure that actions are progressing according to plan, and to guarantee that adequate resources are applied to meet the requirements. These issues are based on a plan–do–study (check)–act cycle, which uses these elements to implement change within the processes of the organization in order to drive and maintain improvements within the processes (see Table 1.3).

The benefits of OHSAS 18001 cannot be overstated; companies large and small have used this standard to great effect, as mentioned above. Here are just a few of these benefits:

- *Improve your image and credibility*: By assuring customers that you have a commitment to demonstrable management of occupational health and safety, you can enhance your image and market share through maintaining a good public image and improved community relations.
- *Improve cost control*: One improvement that all companies are looking for is reduction of costs. The OH&SMS can help with this by increasing rating at insurance companies, while reducing occupational health and safety incidents that may lead to lawsuits and deterioration of the company's image.
- *Use evidence-based decision-making*: By ensuring that you are using accurate data to make your decisions on what to improve, you can greatly increase the chances that your improvements will be successful the first time, rather than having several unsuccessful attempts. By using this data to track your progress, you can correct these improvement initiatives before they go "off the rails," which can save costs and time.
- *Create a culture of continual improvement*: With continual improvement, you can work towards better processes and reduced occupational health and safety hazards in a systematic way in order to improve your public image and potentially reduce your costs, as identified above. When a culture of improvement is created, people are always looking for ways to make their processes better, which makes maintaining the OH&SMS easier.
- *Engage your people*: Given a choice between working for a company that shows care and concern for occupational health and safety and one that does not, most people would prefer the first company. By engaging your employees in a group effort to reduce your occupational health and safety hazards, you can increase employee focus and retention.

So, specifically the advantages of an effective OHSAS management system may be summarized as a system that

---

TABLE 1.3

The PDS(C)A Model Relative to the Standard

Classic Deming Cycle	Standard Clause
Plan	4.1; 4.2; 4.3
Do	4.4
Study (check)	4.5
Act	4.6

- Provides a structured approach for managing OH&S
- Establishes and maintains a commitment to occupational health and safety
- Demonstrates strong commitment to safety excellence
- Organizational structures in place with clear roles and responsibilities
- Existence of a continuous improvement culture
- Strong levels of trust and communication
- Reduction in incident levels with increased measures of performance
- Contributes to business performance by reducing cost and liabilities.

Given the benefits of the OHSAS 18001, is it worth being certified? The answer is, yes. However, in reality it is a decision that an individual organization must make for itself. There are two possible types of certification: (a) certification of a company's occupational health & safety management system against the OHSAS 18001 requirements and (b) certification of individuals to be able to audit against the OHSAS 18001 requirements. In this section, we discuss some approaches (steps) that a company needs to implement an OHSAS 18001 occupational health & safety management system before it receives its certification. This *certification* involves implementing an **occupational health & safety management systems** (OH&SMS) based on the OHSAS 18001 requirements and then hiring a recognized certification body to audit and approve your OH&SMS as meeting the requirements of the OHSAS 18001 standard.

Starting with management support and identifying the legal requirements for the OH&SMS, you will need to start with defining your occupational health and safety (OH&S) policy, occupational health and safety hazards, and OH&S objectives and targets, which together define the overall scope and implementation of the occupational health & safety management system. Along with these, you will need to create the mandatory and additional processes and procedures necessary for your organization's operations. There are several mandatory processes that need to be included and others to be added as the company finds them necessary. For a good explanation on this, take a look at this white paper on *Checklist of Mandatory Documentation Required by OHSAS 18001:2007.* Retrieved on December 15, 2019.

This creation of documents and records can be done internally by your company, or you can get help through hiring a consultant or purchasing standard documentation. To see samples of documentation, visit this *free OHSAS 18001 downloads page.* Retrieved on December 15, 2019.

Once all of the processes and procedures are in place, you will need to operate the OH&SMS for a period of time. By doing this, you will be able to collect the records necessary to go to the next steps: auditing and reviewing your system and becoming certified.

Mandatory Documents
- Scope of the OH&S management system: 4.1, 4.4.4
- OH&S policy: 4.2, 4.4.4
- OH&S objectives and program(s): 4.3.3, 4.4.4
- Roles, responsibilities, and authorities: 4.4.1

- Communication from external parties: 4.4.3.1
- OH&S management system elements & their interaction: 4.4.4
- Operational control procedures: 4.4.6.

Mandatory Records
- Competence, awareness, and training records: 4.4.2
- Record of hazard identification: 4.3.1
- Risk assessment, significance, and controls: 4.3.1
- Performance monitoring records: 4.5.1
- Calibration records: 4.5.1
- Evaluation of compliance records: 4.5.2.1, 4.5.2.2
- Nonconformity, corrective, and preventive action records: 4.5.3
- Internal audit records: 4.5.5
- Management review records: 4.6.

These are the documents and records that are *required* to be maintained for the OHSAS 18001 management system. However, you should also maintain any other records that you have identified as necessary to ensure your management system can function, be maintained, and improve over time.

Additional Documentation   All "real-world" systems require more than the above minimum mandatory requirements to ensure a robust and reliable system. There are several additional documented procedures which are commonly employed to ensure a robust and reliable system, including

- Procedure for hazard identification, risk assessment, and determining controls: 4.3.1
- Procedure for nonconformity, corrective action, and preventive action: 4.5.3
- Procedure for monitoring and measurement: 4.5.1
- Procedure for legal and other compliance requirements: 4.3.2
- Procedure for competence, training, and awareness: 4.4.2
- Procedure for evaluation of compliance: 4.5.2
- Procedure for control of documents and records: 4.4.5, 4.5.4
- Procedure for internal audit: 4.5.5
- Procedure for emergency preparedness and response: 4.4.7
- Procedure for communication, participation, and consultation: 4.4.3.

In the final analysis, to determine what additional documentation you require, simply ask yourself the question: do we need a documented procedure to ensure consistency between employees?

A good maxim to follow is that simple, clear, documentation is more effective than complicated documentation in ensuring that all employees can deliver repeatable outcomes. And don't forget, procedures can be represented in the form of text, flow-chart, pictures, etc. – choose the most effective!

Mandatory Steps to Finish Implementation and Get Your Company Certified

After finishing all your documentation and implementing it, your organization also needs to perform these steps to ensure a successful certification:

- *Internal audit*: The internal audit is in place for you to check your OH&SMS processes. The goal is to ensure that records are in place to confirm compliance of the processes and to find problems and weaknesses that would otherwise stay hidden.
- *Management review*: This is a formal review by your management to evaluate the relevant facts about the management system processes in order to make appropriate decisions and assign resources.
- *Corrective actions*: Following the internal audit and management review, you need to correct the root cause of any identified problems and document how they were resolved.

The company certification process is divided into two stages:

1. *Stage one (documentation review)*: The auditors from your chosen certification body will check to ensure your documentation meets the requirements of OHSAS 18001.
2. *Stage two (main audit)*: Here, the certification body auditors will check whether your actual activities are compliant with both OHSAS 18001 and your own documentation by reviewing documents, records, and company practices.

What OHSAS 18001 Training and Certification
Are Available If You're an Individual?

Training in the concepts of OHSAS 18001 is available, and there are a range of course options for individuals to choose from. Only the first of the courses mentioned below can lead to certification for the individual to be able to audit for a certification body, but the others are very useful for those who will be using these skills within their own company:

- *OHSAS 18001 lead auditor course*: This is a 4- to 5-day training course focused on understanding the OHSAS 18001 OH&SMS standard and being able to use it for auditing management systems against these requirements. The course includes an exam at the end to verify knowledge and competence, and it is only with an accredited course that an individual can become approved to audit for a certification body.
- *OHSAS 18001 internal auditor course*: This is commonly a 2- or 3-day course that is based on the lead auditor course above, but does not include the test for competence, so this is most useful for someone beginning to do internal audits within a company.

- *OHSAS 18001 awareness and implementation course*: Several courses are offered that provide knowledge of OHSAS 18001 and how to implement it. These can be 1- or 2- or even 5-day courses and can even include online e-learning sessions as a method of teaching the material. These courses are good for those who need an overview of the OHSAS 18001 standard, or those who will be involved in the implementation within a company, and many are more economical than investing in the lead auditor course for those involved at this level.

There are a number of accredited training organizations around the world where you can gain individual qualifications in OHSAS 18001. However, because this standard is going to be migrating to ISO 45001 by March 2021, we suggest the effort should be placed on the new standard. It is this reason we do not review the entire standard. Instead, we review its replacement the *ISO 45001*.

This section of the chapter is primarily based on information adapted from the following sites:

- https://www.bing.com/images/search?q=IV+Documentation+Example&FORM=IRMHRS. Retrieved on January 20, 2020.
- https://klariti.com/2015/04/26/review-checklist/. Retrieved on January 20, 2020.
- https://advisera.com/45001academy/what-is-ohsas-18001/. Retrieved on January 20, 2020.
- https://www.doxonomy.com/blog/mandatory-documents-and-records-for-ohsas-18001.
- https://www.slideshare.net/khanhtuyen/checklist-of-mandatorydocumentationrequiredbyiso45001en. Retrieved on January 20, 2020.

ISO 45001

General Comments

ISO 45001 is an ISO standard for management systems of occupational health and safety (OH&S), published in March 2018. The goal of ISO 45001 is the reduction of occupational injuries and diseases (https://www.iso.org/iso-45001-occupational-health-and-safety.html. Retrieved on February 12, 2020).

The standard is based on OHSAS 18001, conventions and guidelines of the International Labour Organization (ILO) including ILO OSH 2001, and national standards (http://www.ilo.org/safework/info/standards-and-instruments/WCMS_107727/lang--en/index.htm. Retrieved on February 12, 2020; https://www.iso.org/iso-45001-occupational-health-and-safety.html. Retrieved on February 12, 2020). It includes elements that are additional to BS OHSAS 18001 (refer to the "ISO 45001 Changes Compared to OHSAS 18001:2007" section below) which it is replacing over a 3-year migration period from 2018 to 2021 (https://www.iso.org/iso-45001-occupational-health-and-safety.html. Retrieved on February 12, 2020). ISO 45001 also follows the HLS of other ISO standards like ISO 9001:2015 and ISO

14001:2015 which makes integration of these standards much easier (https://www.iso.org/publication/PUB100427.html. Retrieved on February 12, 2020; https://auditortraining.pwc.com.au/new-iso-high-level-structure-mean/. Retrieved on February 12, 2020).

While both standards are targeted towards improvements in working conditions, ISO 45001 takes a proactive approach to risk control that starts with the incorporation of health and safety in the overall management system of the organization, thus driving top management to have a stronger leadership role in the safety and health program. On the other hand, OHSAS 18001 takes a reactive approach of hazard control by delegation of hazard control responsibilities to safety management personnel rather than integrating the responsibilities into the overall management system of the company. For more concrete information, see pages 53–54 in the previous section.

ISO 45001 Changes Compared to OHSAS 18001:2007

Context of the organization (Clause 4.1): The organization shall determine internal and external issues that are relevant to its purpose and that affect its ability to achieve the intended outcome(s) of its OH&S management.

What Will Be the Benefits of Using ISO 45001?

An ISO 45001-based OH&S management system will enable an organization to improve its OH&S performance by

- Developing and implementing an OH&S policy and OH&S objectives
- Establishing systematic processes which consider its "context" and which take into account its risks, opportunities, and legal and other requirements
- Determining the hazards and OH&S risks associated with its activities, seeking to eliminate them, or putting in controls to minimize their potential effects
- Establishing operational controls to manage its OH&S risks and its legal and other requirements
- Increasing awareness of its OH&S risks
- Evaluating its OH&S performance and seeking to improve it, through taking appropriate actions
- Ensuring workers take an active role in OH&S matters
- In combination, these measures will ensure that an organization's reputation as a safe place to work will be promoted and can have more direct benefits, such as
  - Improving its ability to respond to regulatory compliance issues
  - Reducing the overall costs of incidents
  - Reducing downtime and the costs of disruption to operations
  - Reducing the cost of insurance premiums
  - Reducing absenteeism and employee turnover rates.
- Being recognized for having achieved an international benchmark (which may in turn influence customers who are concerned about their social responsibilities).

Who Are the Intended Users of the Standard?

The simple answer is all organizations. As long as your organization has people working on its behalf, or who may be affected by its activities, then using a systematic approach to managing health and safety will bring benefits to it. The standard can be used by small low-risk operations equally as well as by high-risk and large complex organizations. While the standard requires that OH&S risks are addressed and controlled, it also takes a risk-based approach to the OH&S management system itself, to ensure (a) that it is effective and (b) being improved to meet an organization's ever changing "context." This risk-based approach is consistent with the way organizations manage their other "business" risks and hence encourages the integration of the standard's requirements into organizations' overall management processes.

Things That an Auditor Should Be Concerned With

Any standard requires some key issues that have to be audited for validation and confirmation as to whether they are being performed and monitor for acceptability and continuous improvement. The ISO 45001 is no different.

How Does ISO 45001 Relate to Other Standards?

ISO 45001 follows the HLS approach that is being applied to other ISO management system standards, such as ISO 9001 (quality) and ISO 14001 (environment). In developing the standard, consideration has been given to the content of other international standards (such as OHSAS 18001 or the International Labour Organization's "ILO – OSH Guidelines") and national standards, as well as to the ILO's International Labour Standards and conventions (ILSs).

Those adopting the standard should find its requirements consistent with the other standards. This will allow for a relatively easy migration from using an existing OH&S management system standard to using ISO 45001 and will also allow for the alignment and integration with the requirements of other ISO management system standards into their organization's overall management processes.

The ISO 45001 has taken a global approach to occupational health and safety management. It is the first OHS standard to be recognized globally. It seeks to create safer workplaces in part by putting a great emphasis on worker and top management participation in safety. As a result, the auditor must be aware of at least the following:

1. The globalization of its use is unique and needs to be understood as a holistic approach to managing an organization.
2. It replaces the ISO 18001 as it emphasizes and integrates much more occupational health and safety issues and concerns into related organizational systems.
3. It provides focus on environment performance and measurement.
4. It requires much greater commitment from the management leadership.
5. It increases the opportunity for proper alignments with strategic goals.
6. It provides protection to the environment by focusing on sustainability and proactive initiatives.

7. It provides more effective communication (horizontal and vertical) through a standardized communication strategy.
8. It focuses on life cycle thinking rather than the immediate result.

Therefore, a good auditor must investigate and understand the value of the standard in providing access to both horizontal and vertical worker participation as a priority. This means that occupational health and safety must not be controlled by "certain" individuals in the organization. This means that attitude and behavior must change where management incorporates this paradigm shift in the entire organization. Words alone will not improve anything. What is needed is integrity, trust, and actions that support the words. To facilitate this paradigm change, organizations will need to set aside adequate resources for worker participation and training on things such as accurate incident reporting, investigations, risk assessment, and other tasks that were the exclusive domain of management under the old system.

Any auditor that has some experience with the ISO standards will tell you that all of them to some degree talk about continuous improvement (CI). In fact, one may say that the concept of CI is a fundamental pillar of these standards. In this line of thinking, it is imperative to understand that in the ISO 45001, we find a tremendous language that clarifies the central notion of OHS. That point is well identified by claiming that management *must identify and respond to nonconformity with action*. This revolutionary idea is projected to something else which is just as powerful. That is, the standard appears that is dismissing the idea of preventive action as a distinct concept. However, that is not quite true. If anything, the standard reinforces the prevention concept by making it a fundamental requirement of the system in its entirety. Very powerful paradigm change! By doing so, it breaks the tradition of reacting to problems with some kind of sorting or other evaluating method. It forces the organization to fix and prevent similar or same problems from happening again.

It is also very interesting that in the ISO 45001, we find the deliberate usage of the term "continual improvement" over "continuous improvement." According to ISO 45001:2018, *continuous* indicates duration without interruption, while *continual* indicates duration that occurs over a period of time with intervals of interruption. Continual seems more suitable for a system that the intent is to safeguard employees from injury and illness, since these processes are implemented before they are evaluated under the plan–do–study (check)–act cycle.

The main difference between the two standards is that ISO 45001 concentrates on the interaction between an organization and its business environment, emphasizing and honing on risks (prevention approach), while OHSAS 18001 was focused on managing OH&S hazards and other internal issues, after the fact of occurrence, i.e., focusing on detection. In addition, ISO 45001 is also process-based versus procedure-based, considers opportunities as well as risks, and includes the views of all interested parties (all the individuals and organizations that can affect your business activities). So, the beautiful approach of the OHS with the new standard is that OHS is no longer separate from everything else an organization does. As an auditor this is a fruitful ground of investigation and probing to find out how committed is management to OHS. The standard gives us the opportunity to do just that in clause 5.1 which requires top management to take "overall responsibility and accountability" for OH&S activities. In addition, the introduction identifies integration of the OHS management system with organizational business factors as a "success factor."

The auditors have a difficult task in here because they have to question the management in a way that the answers are not a "yes" or "no" response. They must uncover what does management do to provide a healthy and safe work environment for all – employees as well as visitors. It is the auditor's responsibility to probe deep with question that will uncover the factors (maybe even policies) that might cause illness, injury, and in some cases death, by mitigating adverse effects on the physical, mental, and cognitive condition of a person. Not surprisingly, the ISO 45001 covers all of those aspects.

Finally, like ISO 14001, in total the ISO 45001 covers ten categories, of which 1–3 are not certifiable. They are informational only. Categories 4–10 are the certifiable clauses, and they are discussed next.

4. *Context of the organization*: ISO now wants you to determine the issues that influence your organization, be they internal or external. External issues will include such things as legal, technological, or cultural, and may be international, national, or local. Internal will include things like values, culture, and knowledge. The needs of interested parties are also to be understood along with the scope of the management system, i.e., what is it covering. Processes, along with their inputs and outputs are to be identified, and documented information will be required as appropriate.

There are two clauses (4.1 and 4.2) which the organization and the auditor must understand because they deal specifically with the needs and expectations of interested parties. Together, these clauses require the organization to determine the issues and requirements that can impact on the planning of the management system and the auditor to ask the appropriate and applicable questions to uncover the effectiveness of the system. As for the external issues, the concerns are as follows:

- The cultural, social, political, legal, financial, technological, economic, natural surroundings, and market competition, whether international, national, or local
- Introduction of new competitors, new technologies, new laws, and the emergence of new occupations
- Key drivers and trends relevant to the industry or sector having impact on the objectives of the organization
- Relationships with, and perceptions and values of, its external interested parties
- Changes in relation to any of the above.

On the other hand, for the internal issues, the concerns are as follows:

- Governance, organizational structure, culture, roles, and accountabilities
- Policies and objectives, and the strategies that in place to achieve them

- The capabilities of the organization understood in terms of resources and knowledge (e.g., capital, time, people, processes, systems, and technologies)
- Information systems, information flows, and decision-making processes (both formal and informal)
- Introduction of new products and equipment
- Relationships with, and perceptions and values of, its internal interested parties
- Standards, guidelines, and models adopted by the organization
- Form and extent of contractual relationships
- Changes in relation to working time requirements and any of the above.

The aim here is to assist the organization in determining its risks, developing or enhancing its policies, setting its objectives, and determining its approach to maintaining compliance with its customers, community, and applicable legal and other requirements. *Note: There is no requirement in the standards to consider interested parties which have been determined not to be relevant. Similarly, there is no requirement to address a particular requirement of a relevant interested party if the requirement is not relevant. Determining what is relevant or not is dependent on whether or not it has an impact on the organization's ability to consistently meet its objectives.*

5. *Leadership*: Top management has to demonstrate leadership. To do this, they need to establish policies and ensure responsibilities, and authorities are communicated and understood. Management also has to promote the discipline across the organization, whether it is quality, environment, or OHS.

6. *Planning*: Organizations now need to use a risk-based approach to address threats and opportunities, and to ensure the management system actually does what it is required to do – that it can prevent or reduce undesired affects and achieve improvement. Objectives and plans need to be developed to meet these objectives; these need to be cascaded through the organization and include responsibilities and time frames. Additionally, changes need to be planned, and the potential consequences (positive or negative) of any change needs to be known. [A reminder is worth repeating here to show the development of identifying risk. The *ISO/DIS 14001:2014 was the only standard to correctly use the terms risk, threat, and opportunity; both ISO/DIS 9001:2014 and ISO/DIS 45001:2014 did not use the term "threat" and therefore suggest that all risk is negative. ISO 45001:2015 clause 6.1.2 does indeed cover the issue of risk. Also, ISO 31000:2009 Risk Management – Principles and Guidelines have been developed to assist organizations in managing risk, however there is no current requirement in the HLS for organizations to specifically use the ISO 31000:2009 format.*]

7. *Support*: Resources need to be provided to support the management system, including providing competent people, appropriately maintained infrastructure and environment, and monitoring and measuring equipment and its calibration. Additionally, the knowledge necessary for the discipline is to be determined, maintained, and made available. The previous document control and records management have been replaced with documented information, where the organization determines what documentation is necessary and the most appropriate medium for that documentation.

8. *Operation*: This replaces Product Realization, Operational Control, and Hazard Identification, Risk Assessment, and Control of Risks in ISO 9001, ISO 14001, and AS/NZS 4801, respectively. There is a stronger emphasis on organizations determining the processes required for their operations, along with appropriate acceptance criteria and contingency plans, e.g., nonconformances, incidents, and emergency preparedness. The HLS also has requirements now for change management and control of external providers (such as contractors, outsourced processes, and procurement).

9. *Performance evaluation*: Performance evaluation takes over from evaluation, data analysis, and monitoring and measurement clauses. Specifically, ISO 14001 and ISO 45001 require an Evaluation of Compliance (Legal and other), while ISO 9001 requires the monitoring of customer satisfaction. Internal audits and management reviews are also included here.

10. *Improvement*: Organizations are required to react appropriately to nonconformities and incidents, and take action to control, correct, deal with consequences, and eliminate the cause so that it does not recur or occur elsewhere. The organization is also required to improve the suitability, adequacy, and effectiveness of the management system. Preventive action is gone – replaced by the risk-based process approach in Section 4 and actions to address risks in Section 6.

Mandatory Documents

- Scope of the OH&S management system (clause 4.3)
- OH&S policy (clause 5.2)
- Responsibilities and authorities within OH&SMS (clause 5.3)
- OH&S process for addressing risks and opportunities (clause 6.1.1)
- Methodology and criteria for assessment of OH&S risks (clause 6.1.2.2)
- OH&S objectives and plans for achieving them (clause 6.2.2)
- Emergency preparedness and response process (clause 8.2).

Mandatory Records

- OH&S risks and opportunities and actions for addressing them (clause 6.1.1)
- Legal and other requirements (clause 6.1.3)
- Evidence of competence (clause 7.2)
- Evidence of communications (clause 7.4.1)
- Plans for responding to potential emergency situations (clause 8.2)
- Results on monitoring, measurements, analysis, and performance evaluation (clause 9.1.1)
- Maintenance, calibration, or verification of monitoring equipment (clause 9.1.1)
- Compliance evaluation results (clause 9.1.2)
- Internal audit program (clause 9.2.2)
- Internal audit report (clause 9.2.2)
- Results of management review (clause 9.3)

- Nature of incidents or nonconformities and any subsequent action taken (clause 10.2)
- Results of any action and corrective action, including their effectiveness (clause 10.2)
- Evidence of the results of continual improvement (clause 10.3).

Non-mandatory Documents (Procedures)   There are numerous non-mandatory documents that can be used for ISO 45001 implementation. The following non-mandatory procedures are the most commonly used:

- Procedure for Determining Context of the Organization and Interested Parties (clause 4.1)
- OH&S Manual (clause 4)
- Procedure for Consultation and Participation of Workers (clause 5.4)
- Procedure for Hazard Identification and Assessment (clause 6.1.2.1)
- Procedure for Identification of Legal Requirements (clause 6.1.3)
- Procedure for Communication (clause 7.4.1)
- Procedure for Document and Record Control (clause 7.5)
- Procedure for Operational Planning and Control (clause 8.1)
- Procedure for Change Management (clause 8.1.3)
- Procedure for Monitoring, Measuring and Analysis (clause 9.1.1)
- Procedure for Compliance Evaluation (clause 9.1.2)
- Procedure for Internal Audit (clause 9.2)
- Procedure for Management Review (clause 9.3)
- Procedure for Incident Investigation (clause 10.1)
- Procedure for Management of Nonconformities and Corrective Actions (clause 10.1)
- Procedure for Continual Improvement (clause 10.3).

2 Industrial Standards/ Requirements

OVERVIEW

Whereas the International Standards (ISO series) account for the basic requirements for many organizations regardless of product, service, or size, that is not enough to address specific industry requirements. The solution that remedies this deficiency came with specific industries developing their own requirements to fit the gap. The automotive industry is one of these industries that developed their own additional requirements in what is known as the IATF 16949.

AUTOMOTIVE STANDARD – IATF 16949

IATF 16949 was jointly developed by the International Automotive Task Force (IATF) members and submitted to the International Organization for Standardization (ISO) for approval and publication. It is a replacement of the ISO/TS 16949 which was a technical specification (TS). The new IATF is considered a standard, and it is especially designed for the automotive industry.

One of the key principles of the IATF 16949 QMS is the focus on improving customer satisfaction by identifying and meeting customer requirements and needs. By improving satisfaction, you improve repeat customer business (loyalty).

So, as of this writing the document is a common automotive quality system requirement based on ISO 9001 and customer-specific requirements (CSR) from the automotive sector. IATF 16949 emphasizes *the development of a process-oriented quality management system* that provides for continual improvement, defect prevention, and reduction of variation and waste in the supply chain. The goal is to meet customer requirements efficiently and effectively.

The specific requirements are 1–10 categories. However, categories 1–3 are only informational. The remaining categories 4–10 are the certifiable clauses and are listed here:

4. Context of the Organization
5. Leadership
6. Planning
7. Support
8. Operation
9. Performance evaluation
10. Improvement
    Annex A: Control Plan.

IATF 16949 puts great emphasis on the compliance with requirements of the standard itself, as well as CSRs. It expands upon the requirements of ISO 9001 regarding internal audits by providing more requirements for an internal audit program and adding requirements for quality management system audits, manufacturing process audits and product audits. In addition, it includes (expands the coverage) a fair number of individual requirements of Ford, GM, and FCA.

Considering the complexity of the QMS (Quality Management System) based on IATF 16949 and the complexity of the automotive industry, the standard has divided internal audits into three categories with specific requirements and different scopes, as well as specific sets of competencies necessary to conduct them. These three categories are discussed next.

Quality Management System + Manufacturing Process + Product Audit = IATF 16949

According to IATF 16949, one of the key requirements in the automotive industry is CSRs which guide your company to gaining customer satisfaction. In this chapter, we will summarize the process of how to successfully define, evaluate, and meet this area of the IATF 16949-based QMS by handling and updating the external documentation.

Customer Satisfaction as a Main Goal

It is important to remember that the IATF 16949 is an automotive quality management system where IATF 16949 member OEM (Original Equipment Manufacturer) companies are the owner of the standard. CSRs are one of the applicable requirements with ISO 9001:2015 and IATF 16949:2016. The main goal within the standard is customer satisfaction by continual improvement, defect prevention, and reduction of waste and variation. If customers for an organization are IATF 16949 member OEMs, it is easy to find CSRs. But for other customers, it must be specifically asked if they have any special manual or procedure for their specific requirements. More information on this may be found in Stojanovic's article describing the *Key Benefits of IATF 16949 Implementation*. (Strahinja Stojanovic, https://advisera. com/16949academy/knowledgebase/key-benefits-of-iatf-16949-implementation/. Retrieved on December 5, 2019.)

Elements in establishing QMS with CSRs: During establishment of a quality management system, one of the sources that must be used is CSRs. These requirements are referenced in elements 4.3.2–7.5.1.1.d and 9.2.2.2 of the IATF 16949 standard. The requirements for these elements are as follows:

Customer-specific requirements: According to IATF 16949 requirement 4.3.2, "customer-specific requirements," all CSRs will be evaluated. By "all customer," it refers to all direct customers of the organization. These can be IATF 16949 OEMs, non-IATF 16949 OEMs, and automotive customers in the supply chain referred to as Tier 1, Tier 2, or Tier 3 customers. CRSs are available in the IATF 16949 website, in customer portals, or upon directly asking a customer in order to share with the organization. Keep in mind that, as CSRs are external documents, updates of these documents will also be monitored. [In the next chapter, we will address the issue of CSRs in more detail.]

Quality management system documentation: According to 7.5.1.1, "Quality management system documentation," an organization's quality management system will include a quality manual. Within the quality manual, a document such as a table, a list, or a matrix will be available which indicates where within the organization's quality management system their CSRs are addressed. Like IATF 16949 and ISO 9001 requirements, CRSs are mandatory for an organization. Within their processes and/or other documentation, organizations must describe how they will implement the requirements. So, like reading IATF 16949 elements one by one, all CRSs must be evaluated and included in an organization's system documentation. For more about documentation in IATF 16949, read the article "How to structure IATF 16949:2016 documentation" (Strahinja Stojanovic: https://advisera. com/16949academy/knowledgebase/how-to-structure-iatf-16949-2016-documentation/. Retrieved on December 5, 2019.)

Quality management system audit: According to 9.2.2.2, "Quality management system audit," an organization will audit all quality management system processes, and customer-specific quality management system requirements will also be sampled to verify effective implementation. This means that, if an organization is using any checklist as an audit questionnaire, CRSs must be included in these questionnaires. The table, list, or matrix that is prepared for a quality manual can be a good source for determining which CRSs will be included in which process questionnaires.

The quality management system audit is what can be considered a classic internal audit. The purpose of this audit is to determine the level of compliance of the QMS with the standard and customer-specific quality management system requirements for effective implementation. The entire scope of the quality management system must be covered within a 3-year calendar period, and an annual internal audit program should be developed. The process approach is a mandatory tool for determining compliance. (For more information, see https://advisera.com/16949academy/knowledgebase/five-main-steps-in-an-iatf-169492016-internal-audit/. Retrieved on December 5, 2019.)

Manufacturing process audit: The purpose of the manufacturing process audit is to determine effectiveness and efficiency of the manufacturing process. The standard requires this type of audit to cover all manufacturing processes within a 3-year calendar period. Considering that, in many cases, the manufacturing process begins and ends with CRSs, the approach to conducting the audit is often defined by the customer. In cases when there are no CRSs for the audit process, the organization itself needs to define the approach. A process-based audit means

- The organization needs to define the audit plan that refers to key processes, instead of just the elements of the standard.
- The auditors need to check performance. This is linked to common metrics for the organization and suppliers.
- The auditors should use the checklist to make sure the audit is thoroughly conducted and have checked all of the required items.
- The auditors should use the organization's procedure for conducting the audit.

The manufacturing process audit includes a check to ensure effective implementation of process risk analysis (such as Failure Mode and Effect Analysis or FMEA), control plan, and associated documents.

Product audit: Products are the ultimate result of the organization's processes. The ability to manufacture products which are compliant with customer and organizational requirements is what QMS is all about. We don't need to discuss how important the product quality is for the automotive industry. Just remember that, in most cases, the majority of the parts installed in the car are not produced by the car manufacturer, but by suppliers. Such great dependence on the suppliers requires thorough audits of the product quality, not only by the company that assembles all the parts but also by the parts suppliers themselves.

The products need to be audited using CSRs at appropriate stages of production and delivery. This allows an organization to verify conformity to their specified requirements. Where not defined by the customers, the organization itself needs to define an approach to be used during the product audit.

Why it has to be done this way: At first glance, the internal audit probably looks like an annoying expense. However, internal audits enable you to discover problems (i.e., nonconformities) that would otherwise stay hidden and cause harm to your business. Let's be realistic. It is human nature to make mistakes, so it's impossible to have a perfect system. It is possible, though, to have a system that improves itself and learns from its mistakes.

In a QMS audit process, it is important to make sure you do not miss anything important, such as *auditor knowledge and proper audit planning*. Keeping this in mind, using a proven method to set up your process can help to greatly simplify implementation. You can use the internal audit process to focus on process improvement instead of just maintaining compliance. By doing this, the company can get more value out of the audits. Process improvement is one of the key elements of an IATF 16949 Quality Management System, so it should be one of the main motivators of a company that wants a strong QMS. Process improvement not only helps with efficiency, but saves time and money in the process. If used properly, the internal audit can be one of the biggest contributors towards QMS improvement instead of being a "necessary evil."

So, let us see what are the specific certifiable requirements of the IATF 16949:

4 Context of the organization
    4.1 Understanding the organization and its context
    4.2 Understanding the needs and expectations of interested parties
    4.3 Determining the scope of the quality management system
    4.3.1 Determining the scope of the quality management system – supplemental
    4.3.2 Customer-specific requirements
    4.4 Quality management system and its processes
    4.4.1.1 Conformance of products and processes
    4.4.1.2 Product safety
5 Leadership
    5.1 Leadership and commitment
    5.1.1 General
    5.1.1.1 Corporate responsibility
    5.1.1.2 Process effectiveness and efficiency

8.4.2.2 Statutory and regulatory requirements

8.4.2.3 Supplier quality management system development

8.4.2.3.1 Automotive product-related software or automotive products with embedded software

8.4.2.4 Supplier monitoring

8.4.2.4.1 Second-party audits

8.4.2.5 Supplier development

8.4.3 Information for external providers

8.4.3.1 Information for external providers – supplemental

8.5 Production and service provision

8.5.1 Control of production and service provision

8.5.1.1 Control plan

8.5.1.2 Requirements of work instructions

8.5.1.3 Verification of job set ups

8.5.1.4 Verification after shutdown

8.5.1.5 Total productive maintenance

8.5.1.6 Management of production tooling and manufacturing, test, inspection tooling, and equipment

8.5.1.7 Production scheduling

8.5.2 Identification and traceability

8.5.2.1 Identification and traceability – supplemental

8.5.3 Property belonging to customers or external providers

8.5.4 Preservation

8.5.4.1 Preservation – supplemental

8.5.5 Post-delivery activities

8.5.5.1 Feedback of information from service

8.5.5.2 Service agreement with customer

8.5.6 Control of changes

8.5.6.1 Control of changes – supplemental

8.5.6.1.1 Temporary change of process controls

8.6 Release of products and services

8.6.1 Release of products and services – supplemental

8.6.2 Layout inspection and functional testing

8.6.3 Appearance items

8.6.4 Verification and acceptance of conformity of externally provided products and services

8.6.5 Statutory and regulatory conformity

8.6.6 Acceptance criteria

8.7 Control of nonconforming outputs

8.7.1.1 Customer authorization for concession

8.7.1.2 Control of nonconforming product – customer-specified process

8.7.1.3 Control of suspect product

8.7.1.4 Control of reworked product

8.7.1.5 Control of repaired product

8.7.1.6 Customer notification

8.7.1.7 Nonconforming product disposition

9 Performance evaluation

9.1 Monitoring, measurement, analysis, and evaluation

9.1.1 General

9.1.1.1 Monitoring and measurement of manufacturing processes

9.1.1.2 Identification of statistical tools

9.1.1.3 Application of statistical concepts

9.1.2 Customer satisfaction

9.1.2.1 Customer satisfaction – supplemental

9.1.3 Analysis and evaluation

9.1.3.1 Prioritization

9.2 Internal audit

9.2.2.1 Internal audit program

9.2.2.2 Quality management system audit

9.2.2.3 Manufacturing process audit

9.2.2.4 Product audit

9.3 Management review

9.3.1 General

9.3.1.1 Management review – supplemental

9.3.2 Management review inputs

9.3.2.1 Management review inputs – supplemental

9.3.3 Management review outputs

9.3.3.1 Management review outputs – supplemental

10 Improvement

10.1 General

10.2 Nonconformity and corrective action

10.2.3 Problem-solving

10.2.4 Error-proofing

10.2.5 Warranty management systems

10.2.6 Customer complaints and field failure test analysis

10.3 Continual improvement

10.3.1 Continual improvement – supplemental

Annex A

A.1 Structure and terminology

A.2 Products and services

A.3 Understanding the needs and expectations of interested parties

A.4 Risk-based thinking

A.5 Applicability

A.6 Documented information

A.7 Organizational knowledge

A.8 Control of externally provided processes, products, and services

Annex A: Control Plan

A.1 Phases of the control plan

Therefore, the auditor's function is to make sure that all these requirements have been completed and followed by the management and employees of the organization. How do they do that? By generating a check list. That check list focuses on the specificity of the requirement, its effectiveness, and whether or not it meets the customer requirements. Specifically, the IATF has 21 mandatory documents that have to be produced for the auditor. They are

1. Product Safety – 4.4.1.
2. Calibration – 7.1.5.2.1
3. Training – 7.2.1
4. Internal Auditor Competency – 7.2.3
5. Employee Motivation – 7.3.2
6. Engineering Specification – 7.5.3.2
7. Design and Development – 8.3.1.1
8. Special Characteristic Defining – 8.3.3.3.
9. Supplier Selection – 8.4.1.2.
10. Out Sources Process – 8.4.2.1
11. Statutory and Regulatory Requirement – 8.4.2.2
12. Supplier Monitoring – 8.4.2.4
13. Control of Changes – 8.5.6.1.
14. Temporary Change – 8.5.6.1.1
15. Rework Procedure – 8.7.1.4
16. Repaired Procedure – 8.7.1.5
17. Disposition of Nonconforming Products – 8.7.1.7
18. Internal Audit Process – 9.2.2.1
19. Problem-Solving – 10.2.3
20. Error-Proofing – 10.2.4
21. Continual Improvement – 10.3.1

However, because the IATF 16969 is an extension of the ISO 9001, by default it includes all of the mandated ISO requirements to this list. Therefore, the complete mandatory list for both documents and records in the IATF 16949:2016 is

1. Scope of the quality management system (clause 4.3)
2. Documented process for the management of product safety-related products and manufacturing processes (clause 4.4.1.2)
3. Quality policy (clause 5.2)
4. Responsibilities and authorities to ensure that customer requirements are met (clause 5.3.1)
5. Results of risk analysis (clause 6.1.2.1)
6. Preventive action record (clause 6.1.2.2)
7. Contingency plan (clause 6.1.2.3)
8. Quality objectives (clause 6.2)
9. Records of customer acceptance of alternative measurement methods (clause 7.1.5.1.1)
10. Documented process for managing calibration/verification records (clause 7.1.5.2.1)
11. Maintenance and calibration record (clause 7.1.5.2.1)
12. Documented process for identification of training needs including awareness and achieving awareness (clause 7.2.1)

13. Documented process to verify competence of internal auditors (clause 7.2.3)
14. List of qualified internal auditors (clause 7.2.3)
15. Documented information on trainer's competency (clause 7.2.3)
16. Documented information on employee's awareness (clause 7.3.1)
17. Documented process to motivate employees (clause 7.3.2)
18. Quality manual (clause 7.5.1.1)
19. Record retention policy (clause 7.5.3.2.1)
20. Documented process for review, distribution, and implementation of customer engineering standards/specifications (clause 7.5.3.2.2)
21. Registry of customer complaints (clause 8.2)
22. Product/service requirements review records (clause 8.2.3.2)
23. Procedure for design and development (clause 8.3.1.1)
24. Record about design and development outputs review (clause 8.3.2)
25. Documented information on software development capability self-assessment (clause 8.3.2.3)
26. Records about product design and development inputs (clause 8.3.3.1)
27. Records about manufacturing process design input requirements (clause 8.3.3.2)
28. Document a process to identify special characteristics (clause 8.3.3.3)
29. Records of design and development controls (clause 8.3.4)
30. Documented product approval (clause 8.3.4.4)
31. Records of design and development outputs (clause 8.3.5)
32. Manufacturing process design output (clause 8.3.5.2)
33. Design and development changes records (clause 8.3.6)
34. Documented approval or waiver of the customer regarding the changes in design (clause 8.3.6.1)
35. Documented revision level of software and hardware as part of the change record (clause 8.3.6.1)
36. Documented supplier selection process (clause 8.4.1.2)
37. Documented process to identify and control externally provided processes, products, and services (clause 8.4.2.1)
38. Documented process to ensure compliance with statutory and regulatory requirements of purchased processes, products, and services (clause 8.4.2.2)
39. Documented process and criteria for supplier evaluation (clause 8.4.2.4)
40. Records of second-party audit reports (clause 8.4.2.4.1)
41. Characteristics of product to be produced and service to be provided (clause 8.5.1)
42. Control plan (8.5.1.1)
43. Total productive maintenance system (clause 8.5.1.5)
44. Records of traceability (clause 8.5.2.1)
45. Records about customer property (clause 8.5.3)
46. Production/service provision change control records (clause 8.5.6)
47. Documented process to control and react to changes in product realization (clause 8.5.6.1)
48. Documented approval by the customer prior to implementation of the change (clause 8.5.6.1)

49. Documented process for management of the use of alternate control methods (clause 8.5.6.1.1)
50. Record of conformity of product/service with acceptance criteria (clause 8.6)
51. Record of expiration date or quantity authorized under concession (clause 8.7.1.1)
52. Documented process for rework confirmation (clause 8.7.1.4)
53. Record on disposition of reworked product (clause 8.7.1.4)
54. Documented process for repair confirmation (clause 8.7.1.5)
55. Record of customer authorization for concession of the product to be repaired (clause 8.7.1.5)
56. Notification to the customer about the nonconformity (clause 8.7.1.6)
57. Documented process for disposition of nonconforming product (clause 8.7.1.7)
58. Record of nonconforming outputs (clause 8.7.2)
59. Monitoring and measurement results (clause 9.1.1)
60. Internal audit program (clause 9.2)
61. Results of internal audits (clause 9.2)
62. Documented internal audit process (clause 9.2.2.1)
63. Results of the management review (clause 9.3)
64. Action plan when customer performance targets are not met (clause 9.3.3.1)
65. Results of corrective actions (clause 10.1)
66. Documented process for problem-solving (clause 10.2.3)
67. Documented process to determine the use of error-proofing methodologies (clause 10.2.4)
68. Documented process for continual improvement (clause 10.3.1)

VDA 6

Overview

Whereas the IATF 16949 is primarily an American Automotive Standard and can be used worldwide for all auto companies, there is also another industry standard that has surface in Europe and is making a strong inroad in the international automotive world. That standard is the VDA 6, and it stands for Verband der Automobilindustrie. It is an organization that sets a bundle of standards. The complete list is

VDA 6.1: 2016 – QM System Audit – Serial Production
VDA 6.2: 2017 – QM System Audit – Service
VDA 6.3: 2016 – Process Audit. This particular standard defines a process-based audit standard for evaluating and improving controls in a manufacturing organization's new product introduction and manufacturing processes. In this book, we are focused ONLY on this one.
VDA 6.4: 2017 – QM System Audit – Production Equipment

VDA 6.5: 2008 – Product Audit
VDA 6.7: 2012 – Process Audit – Production Equipment – Product Creation
Process/Unit Production

For our purposes here, we are only interested in the VDA 6.3 which covers the following items relevant to a complete understanding of what an audit is and what is expected from the audit and auditors. The complete list is as follows:

- Introduction for use which defines the process audit; where it is applicable; classification of audits; and identification of process risks
- Requirements of the auditors which define what exactly must be the credentials of an auditor and what is the code of conduct for auditors
- Audit process which defines the steps and order of an audit
- Potential analysis which define the overall analysis as well as the operational and evaluation analysis
- Evaluating a process audit for material products which elaborates on rules and questions about the audit
- Questionnaire which defines the questionnaire and then elaborates on the specific areas of concern that the questions should address
- Product audit services which delineates a plethora of generic questions in reference to the audit
- Assessment forms which provide a variety sample of forms to be used in the audit
- Best practice/lesson learned which provide a sample of self-assessment and knowledge database
- Definition of terms and glossary which identifies the jargon used, as well as providing a list of abbreviations used in the audit.

From an auditor's perspective, the following are of great importance as they are all touch on the necessary sectors of the production:

- Assessment of product suitability according to Formula Q-capability/Q-capability software.
- Internal audits according to IATF 16949:2016 Formula Q-capability or KVP Continual Improvement Process.
- Self-assessment at least once annually according to Q-capability with documentation.
- Ensure that all risks in the supply chain are clearly identified and systematic measures are implemented.

In addition, they include among others the following important areas:

- Documentation on the development of effective production systems in automobile manufacturing and procedures of supply and distribution.
- Quality assurance measures and responsibility of automobile manufacturers for supply.
- Informative and well-experienced staff and employees in the organization.

The few requirements of the VDA 6.3 certification have all translated into very fruitful benefits. Internationally binding benchmarks have presented the initiative for companies in the automobile sectors towards quality management practices.

VDA 6.3 follows a scoring system, with a provision for automatic disqualification if minimum requirements are not met in process control activities. It is the auditor's responsibility to be aware of these disqualifications and act appropriately.

A CURSORY VIEW OF THE VDA 6.3 THIRD ED. 2016

The Audit Standard VDA 6.3 is a standardized procedure for the conduct of process audits to assess organizational performance and capability for their product realization processes.

The standard is intended to complement and reinforce widely adopted standards such as ISO 9001, IATF 16949, and other sector-specific quality derivatives of ISO 9001 such as those for aerospace and medical equipment. Such audits are a vital element in determining whether an organization/supplier has the capability to meet the needs of a purchaser. Getting this judgment right can make all the difference to the success of a project. Applying the VDA 6.3 audit will support organizations in meeting the IATF requirements regarding manufacturing process audit.

The process approach underpins a vast amount of contemporary business management and strategy whether this is the improvement of existing business processes or the introduction of new processes in support of innovative products and/or services.

So, the Audit Standard VDA 6.3 is a standardized procedure for the conduct of process audits to assess organizational performance and capability for their product realization processes. The standard is intended to complement and reinforce widely adopted standards such as ISO 9001, IATF 16949, and other sector-specific quality derivatives of ISO 9001 such as those for aerospace and medical equipment. Such audits are a vital element in determining whether an organization or supplier has the capability to meet the needs of a customer. Getting this judgment right can make all the difference to the success of the organization accomplishing a *culture of quality*.

VDA 6.3 may also be used to good effect internally within an organization. In this case, the goal of the audit is to analyze processes throughout the whole product life cycle so that risks and weaknesses are detected both in the processes themselves and in their interfaces. The audit report can then form the basis of an improvement plan.

Even though the VDA 6.3 was developed by the Verband der Automobilindustrie (VDA) which is the national association for the German automotive industry, specifically for the German automakers, the last several years the VDA series is being accepted by the international community as a full standard for quality, productivity, *etc*. There is no doubt that the VDA 6.3 is a unique procedure beyond any comparable approach worldwide. It was introduced in 1998, revised in 2010, and substantially overhauled in 2016. The revision has been prompted by an appreciation that the time between the formulation of the concept for a new product and its manufacture is becoming increasingly shorter. As a result, various areas within companies have to operate in parallel to an ever-greater degree which places greater demands on organizations, their processes, and their staff. The standard now covers

- P1: Potential analysis – to assess a new supplier, new location, or new technology
- P2: Project management
- P3: Planning the product and process development
- P4: Carrying out the product and process development
- P5: Supplier management
- P6: Production
- P7: Customer support, customer satisfaction, and customer service.

Risk analysis is a major element of the VDA in the preparation for an audit. Besides the process inputs and outputs, risk analysis must consider the following:

- How does the process operate?
- How is the process supported by various functions and other areas of the organization and by direct personnel?
- What material resources are required to carry out the process?
- What level of effectiveness is being achieved by the process?

The outcome of the VDA 6.3 audit is determined through a quality matrix and is allocated one of three values:

A – Quality-capable
B – Conditionally quality-capable
C – Not quality-capable.

Within the matrix, each of the six elements from project management to customer support (P2P7) is assessed. There are detailed rules for deriving the outcome value from each of the stages. Therefore, VDA 6.3 auditing is a skilled process, and accordingly, VDA 6.3 sets competence requirements for auditors.

For the six process elements, from project management through to customer service, the standard provides a questionnaire which covers the minimum requirements, examples of requirements, and proof of compliance plus reference information. The questions are closed questions for the auditor to answer within the audit, but to get the answers, the auditor must ask open questions, based on the risks identified. [Special note: The VDA 6.3 questionnaire is composed of 58 questions, but the actual number of questions to be answered may be more than 121. The breakdown is as follows: (a) 69 questions if the audit is planned with one manufacturing process step, (b) 95 with two process steps, and (c) 121 with three process steps, which is typical.]

For example, on project management there are seven questions including: Is the project organization established and are tasks and authorities specified for the team leader and team members? The minimum requirements include having a project management process with authorities within the project team clearly specified with all expertise provided for implementation. At the other end of the life cycle, six high-level questions cover customer support, customer satisfaction and service. The first question is "Are the customer's requirements for the Quality Management System, product and process satisfied?". This is a broad question covering, among other things, the organization's quality management certification status, supply of spares for products moving outside series production, packaging meeting customer requirements, product requalification requirements, and conformance with legislation.

The potential analysis is part of the sourcing process and examines potential suppliers. It evaluates the experience of the potential supplier in creating similar products.

In conclusion, VDA 6.3 is an approach that can be used by any manufacturing organization to help drive process improvement. Although developed by the automotive industry, it is not exclusive to automotive organizations.

For individuals with no previous experience, it is strongly recommended that they read/study the actual VDA 6.3 book. In it, they will find out the details of the audit requirements. For example, Module A gives an overview of the requirements of the process audit and its underpinning methodology, which is the best place to start. Module BII is internal auditor training. This module enables delegates to understand the structure, content, and scoring evaluation system of VDA 6.3. They will be able to apply these elements to their own process areas. Module E incorporates Module A, BII, and Module C (the exam module). This qualifies delegates to undertake VDA 6.3 audits externally (at suppliers). For more detailed information, see https://www.industryforum.co.uk/resources/articles/getting-started-vda-6-3-process-audit/. Retrieved on June 9, 2020.

As of this writing, only some customers mandate the use of *VDA 6.3 third edition from 2016*. This current revision makes the standard more precise and easier to adapt it to ever-changing requirements in the automotive industry.

AIAG REQUIREMENTS

OVERVIEW

The Automotive Industry Action Group (AIAG) is a not-for-profit association founded in 1982 and based in Southfield, Michigan. It was originally created to develop recommendations and a framework for the improvement of quality in the North American automotive industry – especially for FCA, Ford, and GM corporations. Its intent was to standardize some of the individual requirements to a global format and make automotive requirements common as much as possible for everyone.

As of this writing, AIAG publishes automotive industry standards and offers educational conferences and training to its members, including the Advanced Product Quality Planning and Control Plan (APQP & CP) and Production Part Approval Process (PPAP), Failure Mode and Effect Analysis (FMEA), Measurement System Analysis (MSA), Statistical Process Control (SPC), quality standards, and reference books. These documents have become *de facto* quality standards in North America that must be complied with by all Tier I suppliers. Increasingly, these suppliers are now requiring complete compliance from their suppliers, so that many Tier II and Tier III automotive suppliers now also comply.

2019 AIAG VERSUS VDA – FMEA

With the proliferation of many recalls in the automotive industry, a global attempt to standardize the FMEA has been an ambition task by many organizations. Ford motor company attempted to standardize the FMEA, but few other companies followed.

This attempt was picked up by the AIAG which formalized – more less – the FMEA for FCA, Ford, and GM. However, there are some differences between the Ford approach and the AIAG which an auditor must know and evaluate appropriately. A comparison is shown in Table 2.1.

In addition to the AIAG, a European organization decided to do the same. That organization is the VDA. So now, there appears to be a standard for the European (VDA) and the AIAG for the US communities. As of this writing, there is a very strong attempt to standardize both these approaches to one but no results yet. Many organizations have adopted the VDA approach, but many have not. Some are still ambivalent about the use of one as opposed the other, and they leave it up to the organization to decide. In any case, for a good short summary of the historical events, see Gruska and Kymal (2019).

TABLE 2.1

A Comparison of the FMEA between AIAG, FCA, Ford, and GM

The Sequence of FMEA

AIAG	FCA	Ford	GM
Generic form	Same	Same	Same
Header	Same	Same	Same
Body of FMEA	Same	Same	Same
Function	Same	Same	Same
Failure	Same	Emphasis is given in identifying at least six failures: 1. No function 2. Unexpected (surprise) failure 3. Degradation over time 4. Intermittent 5. Partial 6. Over-function	Same
Effect	Same	Every effect must be identified and listed separately	Same
Severity	Same	The worst effect is carried over as the numerical severity value. There is only one (1) severity in the column regardless as to how many effects are identified in the previous column. The severity is the highest of the effects	Same
Criticality	Same plus their own	Same plus their own	Same plus their own
Characteristic	Same plus their own	Same plus their own	Same plus their own
Occurrence	Same	Same	Same
Root cause	Same	Same (serious attempts should be made to ID the escape point of the failure)	Same
Current controls	Same	Prevention: Emphasis on the prevention (what have you planned for "catching" the root cause?). It does not affect the detection. Detection: It addresses the "how effective" is the prevention	Same

(Continued)

TABLE 2.1 (*Continued*)

A Comparison of the FMEA between AIAG, FCA, Ford, and GM

The Sequence of FMEA

AIAG	FCA	Ford	GM
Detection	Same	The numerical value for the detection column is the lowest detection number from the previous column – if there are multiple controls for each root cause	Same
Risk Priority Number (RPN)	Same	Ford does not encourage the use of the RPN, even though it appears on the form since it is part of the AIAG format. Instead, Ford emphasizes a three-step approach to define the risk. They are 1. Based on severity 2. Based on criticality 3. Based on the product of severity, occurrence, and detection	Same
Recommended actions	Same	It is strongly recommended that multiple actions be identified so that the "best" solution may be taken	Same
Responsibility and target completion date	Same	Same	Same
Action taken	Same	Action taken must be one of the recommended actions. Generally speaking, Ford is expecting a reduction of occurrence or detection or both. Severity will not change unless a change in design occurs	Same
And appropriate adjustments to severity, occurrence, and detection must be made	Severity may change if the occurrence will change		Severity may change if the occurrence will change
RPN	Same	Same	Same

For the specific FMEA requirements, see Chapter 3.

The two approaches are different, and as a result, a combined FMEA was designed as a hybrid with an official FMEA standard called AIAG & VDA Failure Mode and Effect Analysis – *FMEA Handbook*. The document's intent was to develop a single, robust, accurate, and improved methodology that satisfied both AIAG and VDA FMEA approaches. So, as of this writing, we have the new FMEA which covers the following steps:

- *Step 1*: Planning and Preparation
- *Step 2*: Structure Analysis
- *Step 3*: Function Analysis

- *Step 4*: Failure Analysis
- *Step 5*: Risk Analysis
- *Step 6*: Optimization
- *Step 7*: Results Documentation.

These seven steps are organized into three phases:

- Steps 13 represent the "System Analysis" phase.
- Steps 46 represent the "Failure Analysis and Risk Mitigation" phase.
- Step 7 represents the "Communication and Results Documentation" phase.

Although the new AIAG-VDA method supports an evolution of distributed design and extended supply chains, not everyone is on board with accepting the standard as their own standard. If a supplier wants to use the AIAG-VDA FMEA, it will be accepted, but some OEMs prefer (but will accept) their own until further notice (e.g., Ford Motor Co. prefers their own version (*FMEA Handbook* version 4.2). The plethora of Methodologies; structural changes in the relationships between customer, supplier, and sub-supplier; links between Design Failure Mode and Effect Analysis (DFMEA)/DVPR (test plans); process flow/Process Failure Mode and Effect Analysis (PFMEA); control plans; and shop-floor documents will cause a disruptive, innovative wave – unless they become standardized for everyone. OEMs and automotive software will need to link both requirements flow and failure flow down in this extended, distributed environment. There will be a next level of design as well as quality innovation and improvement with this evolution. With the product shift globally from gas and diesel vehicles to electric and autonomous cars, the AIAG-VDA structure analysis in the DFMEA may be quite timely.

Changes that the auditor must be aware of and be careful in evaluating them:

- The form is different (the new one has more columns).
- The order of failure–effect–cause is redefined to effect–failure–cause. That is a problematic issue because of the possibility of not identifying the appropriate failure.
- The introduction of *Action Priority* (AP) as an ordering of issues that should be addressed to manage risk and is based on an elaborate table that combines severity and occurrence (It is the old way of defining criticality). The AP replaces the Risk Priority Number (RPN)
- Results and documentation are very nebulous, and it presents a subjective approach to evaluating. Auditors must be very vigilant in this area.
- In risk analysis, the VDA handbook makes prevention control an integral part of rating occurrence. It does this by essentially changing the operational definition of occurrence. Occurrence traditionally has been a conditional probability – the probability that a cause would arise and a failure mode (FM) would ensue. Now, the *VDA Handbook* says that this conditional probability must account for the effectiveness of any existing prevention controls. In other words, it's now the probability that a cause will arise, it won't be blocked by a prevention control, and therefore no FM will follow. This forces an implicit assumption about

the effectiveness of the prevention control, something that was explicit in the fourth edition AIAG manual. This may be very problematic for many suppliers as now they are expected to know what will happen in the future with a specific prevention action. (In essence, we are asking the supplier to be a prophetic voice, something that is not humanly possible to do. That will cause problems, if not frustrations between organizations and customers.)

As an auditor, one has to watch and probe for the following:

- First, it can be seen that the form has many more columns than previously.
  - Structural analysis
    - System
    - System element (interface)
    - Component element (item or interface).
  - Functional analysis
    - Function of system and requirements or intended output
    - Function of element and requirements and intended performance output
    - Function of component element and requirement or intended output or characteristic.
  - Failure analysis
    - Effect of failure
    - Severity
    - Failure mode (FM)
    - Cause of failure (FC).
  - Risk analysis
    - Current prevention of control of failure cause (FC)
    - Occurrence
    - Current detection control (FC) or (FM)
    - Detection of FCFM
    - AP (Action Priority)
    - Filter code.
  - Optimization
    - Prevention action
    - Detection action responsible person
    - Target completion date
    - Status (untouched, under construction, in progress, completed, discarded)
    - Action taken with pointer to evidence
    - Completion date
    - Severity
    - Occurrence
    - Detection
    - Action Priority.
- Second, the first three grouped headings, which are structure analysis, function analysis, and failure analysis, are organized in threes, representing the focus element, the higher element, and the lower element.

- The fourth grouped heading is risk analysis, which no longer includes the RPN but introduces the AP step. This is a major inclusion in the new VDA format. It is recognized as FMEA-MSR (Failure Mode and Effect Analysis-Monitoring and System Response) and is intended to maintain a safe state (safety) or state of regulatory (environmental) compliance during customer usage.

The replaced AP is based on the severity, occurrence, and detection ratings. The AP is not a risk priority but rather a priority for action (high, medium, or low) to reduce the risk of failure to function as intended. While the RPN encouraged ranking potential failures from high to low, the elaborate AP tables of the AIAG-VDA define an absolute priority for action regardless of how many other items have been identified in the FMEA study.

- The sixth grouped heading is optimization, which shows the recommended actions planned and the action taken. These are distinctly different items. The preventive (planned) items are what has been planned to avoid the failure of the root cause and the detection action is how effective is the planned action to "catch" the problem. Therefore, the detection numerical value is based on the effectiveness of the detection and NOT on the prevention.
- Although the FMEA process was developed so that teams could use the forms to develop their FMEAs, software to develop FMEAs will become commonplace with the AIAG-VDA FMEA because it represents a three-dimensional relationship not easy to capture in an Excel form.

3 Customer Specific

OVERVIEW

As it was mentioned in the last two chapters, the international as well as the industrial standards quite often do not fulfill all the requirements that "a" customer may have for their own satisfaction from their suppliers. That means that both the international and industrial standards create or perhaps fall short of specificity to provide the expected improvement from their point of view. Therefore, this chapter continues the effort of understanding and closing the "gap" with an overview of the specific requirements for the three major American automotive companies that add completeness to both international and industrial requirements.

FIAT CHRYSLER AUTOMOTIVE (FCA)

FCA's FOREVER REQUIREMENTS

These requirements are a group of activities that organizations must perform to ensure process stability and the quality of the manufactured product. They include, among others,

1. Proactive communication with FCA based on (a) preventive customer information about quality and (b) logistic issues.
2. Approval request from FCA before any accelerated process change begins.
3. Customer approval request before introducing any internal change or change of the Tier 2 production location.

All three of these fundamental requirements are indeed part of the FCA Forever Requirements and are based on the IATF 16949 standard. Specifically,

- *Requirement 7.4. – Communication*: informs business partners that they should comply with the Forever Requirements described in the SQ.00012 procedure. This procedure is currently very important for persons who are working in quality departments and project managers who are in direct contact with FCA SQEs (supplier quality engineers). There is a very helpful checklist in it that has a clear structure informing the organization of which activities based to a large extent on the risk analysis should be considered while submitting the checklist for a given change.
- *Requirement 8.3.1. – Type and extent of control – supplemental*: in relation to its sub-suppliers, the organization should cascade and forward all quality requirements of the FCA group, e.g., quality planning, Production

Demonstration Run (PDR), and Forever Requirements. During the process audits and Advance Quality Planning, this requirement can be checked by FCA SQE during their visit based on the new Production Part Approval Process (PPAP) audit referring to the SQ.00010 standard. The requirement that must be met by the supplier in this point applies to clause 17.79.

- The last customer-specific requirement (CSR), which is indirectly related to Forever Requirements, is included in *requirement 8.5.3 – Property belonging to customers or external providers*, which informs suppliers that increase productivity by duplication of forms, on which approval was already produced product based on the Plant Evaluation after the Pre-Series phase, should be managed through Forever Requirements.

In the case of even partial FCA Forever Requirements, the client's representatives may use a higher escalation tool such as Forever Requirements Violation (FRV). This applies in particular to situations referring to unauthorized changes in the process, construction (design), material, sub-supplier's location/material, and change in tooling capacity. Another issue that may necessitate the use of the FRV is the concern that a supplier may use approved parts from a non-approved location.

FCA's Specific Requirements

The totality of the CSR for the FCA is summarized here based on the internet site of: https://www.iatfglobaloversight.org/wp/wp-content/uploads/2019/07/FCA-US-LLC-CSR-IATF-16949-20190708.pdf. Retrieved on January 10, 2020. Also, https://www.iatfglobaloversight.org/wp/wp-content/uploads/2018/04/FCA-US-LLC-CSR-IATF-16949-20180412.pdf. Retrieved on January 10, 2020.

Clause 4: Context of the Organization

4.1–4.4.1.1 No additional FCA US customer-specific requirement for these sections.

4.4.1.2 *Product safety organizations* seeking assistance with implementing product safety compliance processes for the United States should refer to the document Model Vehicle Safety Compliance Program, found in the "NAFTA References" section on eSupplierConnect. [As of this writing, the United States-Mexico-Canada Agreement (USMCA) has not been implemented yet.]

4.4.2 No additional FCA US customer-specific requirement for this section.

Clause 5: Leadership

5.1–5.3 No additional requirements.

5.3.1 *Organizational roles, responsibilities, and authorities – supplemental*: The organization shall create and maintain records for all applicable professional roles in the FCA application Supplier Information Card (SIC) (7.2.2).

5.3.2 No additional requirements.

Clause 6: Planning for the Quality Management System

6.1–6.3 No additional requirements.

Clause 7: Support

7.1–7.2.1 No additional requirements.

7.2.2 *Competence – on-the-job training*: Each location shall have a sufficient number of trained individuals such that computer applications necessary for direct support of FCA US manufacturing can be accessed during scheduled FCA US operating times, and other applications can be regularly accessed during normal business hours. Where FCA US computer applications are specified for use by more than one organization operational area (e.g., GIM use by both manufacturing and material supply), each area shall have individuals trained and available for direct support of FCA US during scheduled operating times. The specific computer applications required will vary with the scope of an organization site's operations. For manufacturing sites, the recommended applications include, but not limited to,

- 3CPR – Third-Party Containment and Problem Resolution
- beSTandard – FCA Global Standards Database
- CQMS – Corporate Quality Management System
- DRIVe – Delivery Rating Improvement Verification
- EWT – Early Warranty Tracking
- GEBSC – Global External Balanced Scorecard
- GIM – Global Issue Management
- GCS – Global Claims System
- NCT – Non-Conformance Tracking
- PC Portal II – Production Control Portal II
- PRAS – Parts Return Analysis System
- QNA – Quality Narrative Analyzer
- SIC – Supplier Information Card
- webCN – Change Notice System
- WIS – Warranty Information System.

Notes: (a) All applications listed above are accessible through eSupplierConnect (8.2.1.1), and (b) FCA US periodically offers training to organization personnel on selected FCA US processes and procedures (including those referenced in this document), during Supplier Training Week. Information on content, scheduling, and registration is available in the "Supplier Learning Center" application in eSupplierConnect.

7.2.3–7.5.3 No additional requirements.

7.5.3.2.1 *Record retention organization-controlled records*: Records identified in this document as "organization-controlled" shall be retained on-site, but made available for review upon request by FCA US or the Certification Body (CB). Minimum Retention Requirements Retention of Design Verification (DV) and Performance Verification (PV) data, records, and samples shall conform to the requirements of PF-8500. Quality performance records (e.g., control charts, inspections, and test results) shall be retained for one calendar year after the year in which they were created. Records of internal quality system audits and management review shall be retained for 3 years.

7.5.3.2.2 No additional requirements.

Clause 8: Operation

8.1–8.2.1 No additional requirements.

8.2.1.1 *Customer communication – supplemental*: The organization shall establish a connection for electronic communication with FCA US through eSupplierConnect. **Note**: Instructions for registering for the portal and assistance with its use are found at https://fcagroup.esupplierconnect.com.

8.2.2–8.2.2.1 No additional requirements.

8.2.3 Review of requirements related to the products and services

8.2.3.1 *Review of the requirements*: The organization shall conduct a review of the provided Additional Quality Requirements (AQRs) in accordance with SQ.00001 and the Master Process Failure Mode and Effects Analysis (MPFMEA) documents accordance with SQ.00007 prior to responding to any source package tendered by FCA US.

8.2.3.1.1 No additional requirements.

8.2.3.1.2 *Customer-designated special characteristics*: FCA US has identified a series of classifications to identify all characteristics of parts, components, or systems. These are summarized in the following list:

- *Regulatory*: Regulatory characteristics have an impact on the safety or emissions performance of the vehicle or are expected to be important for vehicle homologation.
- *Critical*: Deviation from the required specifications of critical characteristics may compromise the efficiency or use of the product by the customer.
- *Capability*: Deviation from the required specification of capability characteristics may cause potential problems with efficiency, use, or vehicle assembly. These characteristics are used primarily to establish product capability and to aid root cause analysis.
- *Ordinary*: Features affecting the function of the part. Characteristics typed as regulatory, critical, or capability are identified with special symbols on FCA US engineering drawings.
- Use of characteristic classifications for FCA US parts, components, or systems shall conform to CEP12679.

8.2.3.1.3–8.3.2 No additional requirements.

8.3.2.1 *Design and development planning – supplemental:* FCA US uses the Advance Quality Planning and PPAP (documented in SQ.00010 Advance Quality Planning (AQP) and Production Part Approval Process (PPAP)), to identify and manage product development tasks that are critical to quality. When required, organizations shall participate in teams to develop parts or components, and shall use APQP/PPAP. A FCA US-led AQP/PPAP program shall be performed for parts that have a customer-monitored (high or medium) initial risk as identified by the FCA US SOE. Supplier-monitored (low-risk) parts shall have an organization-led program, unless otherwise specified by the SOE. Parts that have been out of production for 12 months or more shall have an organization-led AQP/PPAP unless otherwise determined by the SOE. AQP shall be completed prior to providing Pre-Series (PS) level parts to FCA US and shall be completely approved prior to a PPAP submission. In the event that use of AQP/PPAP is not required, organizations shall develop products according to the Advanced Product Quality Planning (APQP) process.

8.3.2.2–8.3.3.1 No additional requirements.

8.3.3.2 *Manufacturing process design input*: The organization shall include AQR and MPFMEA provided by FCA US as inputs to manufacturing process design.

8.3.3.3 *Special characteristics*: The organization shall document the equivalence of the internal special characteristic symbols with FCA US equivalent symbols and refer the equivalence when the organization uses internal symbols in its communications with FCA US.

8.3.4–8.3.4.1 No additional requirements.

8.3.4.2 *Design and development validation*: DV and PV shall be conducted in conformance with PF8500. DV and PV shall be satisfactorily completed before AQP and PPAP approval. **Note:** Guidance on the extent of required PV testing is provided by the PPR/PA tool Production Validation Testing Scope.

8.3.4.3 No additional requirements.

8.3.4.4 *Product approval process audit*: A systematic and sequential review of the organization's process shall be completed through a process audit (PA) performed by the FCA SOE and product engineer prior to a PPAP submittal. The purpose is to verify the organization's process readiness and to assure understanding of complete program requirements. The organization shall comply with Production Part Approval Process (PPAP), 4th Edition, Service Production Part Approval Process (Service PPAP), 1st Edition, and FCA US Customer-Specific Requirements for Use with PPAP, 4th Edition.

8.3.5 *Design and development outputs organizations preparing*: DFMEAs should follow the AIAG Potential Failure Mode and Effects Analysis (FMEA). The AIAG/VDA FMEA, Design FMEA, and Process FMEA Handbook may be used, with analysis documented on the alternate form (Form B).

8.3.5.1 No additional requirements.

8.3.5.2 *Manufacturing process design output*: PFMEAs and control plans are required for prototype, pre-launch, and production phases. PFMEA and control plan documentation shall be audited to the PFMEA and Control Plan Document Audit Form. Control plans shall be verified to the Control Plan Process Audit Checklist, with corrective action for any identified nonconformance(s) documented on the associated PDCA Planning Worksheet. A FCA US representative's signature is not required on control plans, unless specifically requested by the SOE. Organizations preparing PFMEAs should follow the AIAG Potential FMEA. The AIAG/VDA FMEA, Design FMEA, and Process FMEA Handbook may be used, with analysis documented on the alternate form (Form G).

8.3.6–8.4.1.1 No additional requirements.

8.4.1.2 *Supplier selection process*: With respect to suppliers to the organization ("sub-tier suppliers"), the organization shall

- Conduct an on-site process audit (or equivalent) and PDR for all parts/suppliers that are NOT considered by FCA US or the organization to be low risk to the vehicle program.
- Develop and maintain a list of approved suppliers for each sub-component, raw material, commodity, technology, or purchased service that is not consigned or directed by FCA US. The organization shall have a documented process and use assigned personnel to monitor and manage performance.

8.4.1.3 *Customer-selected sources (directed parts/consigned parts)*: For both directed parts and consigned parts, FCA US is responsible for leading the Advance Quality Planning (process planning review for some existing programs), process audit, and PDR activities up to and including PPAP, with input from and participation of the organization. If the organization receives directed parts or materials, the organization is responsible for managing the ongoing quality of the supplier components following PPAP, working with FCA US to resolve issues. If the organization receives consigned parts or materials, FCA US is responsible for managing the ongoing quality of the supplier components following PPAP, with input from and participation of the organization.

8.4.2–8.4.2.1 No additional *requirements.*

8.4.2.2 *Statutory and regulatory requirements:* See Section 4.4.1.2 for guidance on implementing safety compliance processes for the United States.

8.4.2.3 *Supplier quality management system development*: Management of Supplier Quality Management System (QMS). Development supplier QMS development effectiveness shall be evaluated on the basis of evidence that the organization has processes in place that include such elements as

- Supplier QMS development strategy (8.4.2.5), using risk-based thinking to establish
  - Minimum and target development levels for each supplier.
  - Criteria for designating "exempt" suppliers.
  - Criteria for granting waivers to select suppliers for compliance to specified elements of ISO 9001 or IATF 16949.
- Second-party audit administration (8.4.2.4.1).
  - Identification of second-party auditors.
  - Criteria for granting self-certification status to qualified suppliers.
  - A schedule for second-party audits.
- Organization-controlled record keeping (7.5.3.2.1).
- Progress monitoring. At a minimum, the organization shall require their non-exempt suppliers to demonstrate compliance to ISO 9001 and MAQMSR.

Note: Organizations requiring additional guidance on supplier QMS development should refer to CQI19: Sub-tier Supplier Management Process Guideline. Minimum Automotive Quality Management System Requirements for Sub-Tier Suppliers (MAQMSR). The organization shall prioritize the QMS development program for non-exempt suppliers to introduce compliance to the MAQMSR, as the first step beyond compliance with ISO 9001 or certification to ISO 9001. Ship-Direct Suppliers

Organizations may, with FCA US Purchasing concurrence, identify a supplier location within FCA Purchasing systems as an organization manufacturing site. (Such a designation allows direct shipment of manufactured goods to FCA US.) Unless otherwise specified by FCA US, such sites shall be subject to the registration requirements described in Section 1.2. In the event that FCA US chooses to grant such a supplier site an exemption to IATF 16949 registration,

- The site shall receive the highest priority for QMS development.
- The site shall not be designated "exempt," or a "waiver" shall not be granted, without the written concurrence of FCA US supplier operations. Suppliers certified to IATF 16949 Supplier QMS certification by an IATF-recognized CB to IATF 16949, completely satisfy the requirements for quality management system development. Further QMS development by the organization is not required, while the supplier's certification is valid. If the supplier certification expires or is cancelled or withdrawn by their CB, the organization shall establish and implement a plan for second-party audits to ensure continued compliance to IATF 16949 until the supplier is recertified. Exemption shall not be granted as an alternative to recertification without approval from FCA US Supplier Operations management.

8.4.2.3.1–8.4.2.4 No additional requirements.

8.4.2.4.1 *Second-party audits, second-party audit administration*: The second party must annually audit each non-exempt supplier for whom it has performed the second-party service.

- For suppliers not certified to ISO 9001, the duration of these audits must conform to the full application of the audit day requirements of the Rules, Section 5.2.
- For ISO 9001-certified suppliers, audit length may vary to suit individual supplier requirements and audit resource availability in accordance with the documented development strategy. Audit reports shall be retained as organization-controlled records (7.5.3.2.1). The following second-party qualifications shall apply:
  1. The organization must be certified to IATF 16949:2016 by an IATF-recognized CB.
  2. The IATF 16949 certification of the second party cannot be in "suspended" status. Supplier self-certification. If the organization has suppliers for whom self-certification is an effective alternative to second-party audits for QMS development, the organization shall have a documented process for identifying and qualifying self-certifiable suppliers. Qualification criteria shall include a preliminary evaluation (audit) of the supplier's QMS, an analysis of the supplier's quality performance, and an assessment of the incremental risk to organization products. Self-certification qualifications shall be documented and subject to periodic review. Such documents shall be managed as organization-controlled records (7.5.3.2.1). 8.4.2.5 Supplier

development supplier exemptions/waivers. The organization strategy for supplier development of its active suppliers shall include a documented process for designating "exempt" suppliers – those suppliers who are unable or unwilling to fully certify a quality management system to IATF 16949 or ISO 9001. The organization development strategy shall include provisions for granting partial exemptions ("waivers") to suppliers providing commodities for which specific sections of ISO 9001 or IATF 16949 do not apply. Except as noted in Section 8.4.2.3, declaring a supplier as "exempt" does not relieve the organization of the responsibility for supplier QMS development for any sections of ISO 9001 or IATF 16949 not explicitly waived. Supplier development prioritization, exemption, and waiver decisions, as well as the scope of individual exemptions or waivers, shall be documented and subject to periodic review. This documentation shall be retained as an organization-controlled record.

8.4.3 No additional requirements.

8.4.3.1 *Information for external providers – supplemental*: With respect to external providers to the organization (i.e., "sub-tier suppliers"), the organization shall

- Cascade and communicate all FCA US quality requirements (e.g., Quality Planning, Process Audit, PDR, and Forever Requirements) throughout the organization's supply chain.
- Initiate a Forever Requirement Notice for any proposed process change throughout the supply chain.

8.5–8.5.1 No additional requirements

8.5.1.1 *Control plan*: For characteristics identified on the control plan as critical (8.2.3.1.2), the organization shall conduct a monthly dimensional study in accordance with QR-10012 and SPB-00001-09.

8.5.1.2–8.5.2 No additional requirements.

8.5.2.1 *Identification and traceability – supplemental*: Organizations shall conform with PF.901106 when providing parts or components:

- That require tracking to ensure emission, certification, and regulatory compliances.
- That are designated as high-theft components for law enforcement needs.

8.5.3–8.5.4 No additional requirements.

8.5.4.1 *Preservation – supplemental*: Organizations shall be familiar and comply with FCA US packaging, shipping, and labeling requirements contained in the Packaging and Shipping Instructions manual.

8.5.5–8.5.6 No additional requirements.

8.5.6.1 *Control of changes – supplemental*: The organization shall comply with the Forever Requirements activities described in SQ.00012 Forever Requirements.

8.5.6.1.1–8.6.1 No additional requirements.

8.6.2 *Layout inspection and functional testing*: Layout Inspection – Production to ensure continuing conformance to all FCA US requirements, the organization shall implement a program to conduct a complete layout inspection of all organization-manufactured parts and components including all subcomponents. Unless otherwise specified by FCA US Engineering and Supplier Operations, the reference standard for layout inspections shall be the released FCA US Engineering drawing. The approved control plan shall also be used where applicable.

The frequency of layout inspections for production parts and components shall be established following an assessment of risk to product quality. In the absence of risk analysis, the inspections shall be conducted annually. Evaluation of program effectiveness shall be based on evidence that the organization has a process in place that includes elements such as

- An assessment of risk of nonconformance (6.1.1, 6.1.2, 6.1.2.1).
- An established inspection schedule.
- Qualified inspectors identified and employed (7.2.3).
- Conformance evaluation of non-consigned, externally provided subcomponents (8.4.2, 8.4.2.1).
- A defined corrective action process, including (a) customer notification of nonconformance (8.7.1.6), (b) corrective actions (8.7.1), and (c) verification of corrective action effectiveness.
- Record retention (7.5.3.2.1). Inspection frequencies >1 year require a written waiver by FCA US Supplier Operations. Any such waiver shall be subject to annual review and renewal. Documented evidence of the waiver shall be retained as an organization-controlled record. Layout Inspection – Service. The frequency and extent of layout inspections for service parts and components shall be established by the organization with the written approval of Mopar Supplier Quality. Documented evidence of the approved layout inspection plan shall be retained as an organization-controlled record. In the absence of a written agreement, a production-level inspection program (per above) is required.

8.6.3 *Appearance items*: Organizations that provide appearance items – parts or components whose color, gloss, or surface finish requirements are specified by the FCA US Product Design Office – shall conform with AS-10119. The FCA US Product Design Office specifies and controls all appearance masters. Samples of appearance masters are available from the Thierry Corporation: http://www.thierry-corp.com [(248) 549–8600, 49 (0) 711-839974-0.]

8.6.4 No additional requirements.

8.6.5 *Statutory and regulatory conformity*: See Section 4.4.1.2 for guidance on implementing safety compliance processes for the United States.

8.6.6–8.7.1 No additional requirements.

8.7.1.1 *Customer authorization for concession*: The organization shall obtain written approval from FCA US Engineering and Supplier Operations prior to implementing procedures for repair or reuse.

8.7.1.2 *Control of nonconforming product – customer-specified process*: The organization shall use the NCT System as directed by FCA US to manage potentially nonconforming and nonconforming material shipped to FCA US facilities (assembly plants, powertrain plants, stamping plants, Mopar parts depots), as well as Extension of Plant (EOP) operations and Module Suppliers. The organization shall also comply with all applicable process requirements specified in SQN-A0469 Supplier Incident Management – NAFTA. [As of this writing, the requirements for the USMCA have not yet been specified.] When directed by FCA US for the containment of nonconforming material, the organization shall comply with all program policies and project requirements for the 3CPR Web-Based System, as specified in the General Terms and Conditions and documented in SQN-A0489 Third-Party Containment and Problem Resolution (3CPR).

8.7.1.3 *Control of suspect product*: Parts and components marked for obsolescence on a FCA US Engineering CN (change notice) shall be classified and controlled as nonconforming product. The organization shall disposition such parts and components in accordance with Section 8.7.1.7.

8.7.1.4 *Control of reworked product*: The organization shall obtain written approval from FCA US Engineering and Supplier Operations prior to implementing procedures for rework.

8.7.1.5–8.7.1.7 No additional requirements.

Clause 9: Performance Evaluation

9.1–9.1.1.3 No additional requirements.

9.1.2 *Customer satisfaction*: Global External Balanced Scorecard FCA US. Purchasing uses the Global External Balanced Scorecard (GEBSC) to evaluate customer satisfaction with its external production and service (Mopar) suppliers. The production report displays ratings for five Operational Metrics:

- Incoming Material Quality (IMQ).
- Delivery.
- Warranty.
- Cost.
- Overall. The Mopar report displays ratings for three Operational Metrics:
- Incoming Material Quality (IMQ).
- Delivery.
- Overall. The metrics used by FCA US to evaluate the performance of the organization's quality management system are IMQ, Delivery, and Warranty (where applicable). The remaining Operational Metrics and the Strategic Metrics shall not be used.

    Notes: (a) Data for organizations managed by FCA US Purchasing appear in "NAFTA" region-filtered reports [As of yet, there are no specific requirements for the USMCA]. (b) The GEBSC display "By Location/Material Group" evaluates organization site performance at a commodity level. Supplier Quality Reporting FCA US may provide Certification Bodies with periodic reports of their clients' quality data, such as GEBSC

Incoming Material Quality (IMQ), Delivery, and Warranty metrics with supporting data.

- FCA US Supplier Operations process audit reports. **Note**: Sharing CB client quality data does not constitute an OEM performance complaint as described in Section 8.1 of the Rules.

9.1.2.1 *Customer satisfaction – supplemental OEM*: Performance complaint FCA US may file an OEM performance complaint when confronted with a specific organization responsible quality performance issue, where a root cause may be a nonconformance in the organization's quality management system. FCA US shall initiate an OEM performance complaint by sending the appropriate Oversight office a notification letter that will

- Identify the organization site and their CB.
- Summarize substance of the complaint.
- Document the affected element(s) of IATF 16949.
- Request a copy of the organization site's last audit report. **Note**: As FCA US is an IATF member; a request for client audit reports is permitted under Section 3.1.e of the Rules.
- Request the Oversight office witness the Special Audit conducted to verify implementation of corrective action. Upon receipt of the OEM performance complaint notification letter from the Oversight office, the CB shall investigate the complaint in accordance with Section 8.0 of the Rules. An OEM performance complaint may be filed in conjunction with, or independently of, a TPSL action. The CB findings from an OEM complaint investigation may be used by FCA US to establish the need to place an organization site in TPSL or New Business Hold. Top Problem Supplier Location (TPSL) Reporting Upon periodic review of EBSC quality measures and other key performance indicators, FCA US may notify specific organization sites that they have been identified as a Top Problem Supplier Location (TPSL). The TPSL designation signals FCA US dissatisfaction with the organization site's quality performance, and begins a process to develop and implement a performance improvement plan. FCA US shall notify the CB of the organization site's involvement in the TPSL process by sending the CB a copy of the notification letter and follow-up communications (as required) that will
  - Identify the organization site.
  - Summarize the process.
  - Document specific areas of concern, with supporting data.
  - Request a copy of the organization site's last audit. **Note**: As FCA US is an IATF member, a request for client audit reports is permitted under Section 3.1.e of the Rules. CB notification of TPSL activity is for information only and does not constitute an OEM performance complaint as described in Section 8.1 of the Rules. However, FCA US reserves the right to file a performance complaint at any point within the TPSL process. FCA US shall notify the CB when the organization site has

achieved the agreed-upon exit criteria and is removed from the TSPL process. Quality New Business Hold (QNBH) upon periodic review of EBSC quality measures and other key performance indicators, FCA US may notify an organization that they have been placed in QNBH status. This indicates that the organization site's quality performance is persistently below expectations and corrective action is required. **Note**: While in QNBH status, the organization will be ineligible to bid on new FCA US business supplied from the affected organization site(s) without Purchasing Senior Management intervention. A notification letter is sent to the organization, outlining the substance of the complaint and identifying the exit criteria the organization must achieve to be removed from QNBH status. FCA US will file an OEM performance complaint in a separate letter sent to the Oversight office of the organization's CB via electronic mail. Upon completion of the process in accordance with Section 8.0 of the Rules, the organization will remain in QNBH status, while FCA US monitors GEBSC quality measures and other key performance indicators. When the QNBH exit criteria established for the organization have been met, FCA US shall

– Remove the QNBH status, lifting the associated commercial and quality sanctions. (Sanctions imposed by other FCA US processes may remain in place.)
– Notify the affected organization site(s), the CB, and the Oversight Office. If the CB withdraws the certificate upon completion of the process in accordance with Section 8.0 of the Rules, FCA US Purchasing management will develop a joint plan for the organization that either restricts further commercial activity or works towards improving processes and performance to a level that supports organization efforts to recertify. If an organization site is seeking certification to IATF 16949, but is placed on QNBH status before the Stage 2 audit is conducted, the CB shall not conduct a Stage 2 audit until the QNBH status is lifted or FCA US Supplier Operations management notifies the organization and the CB in writing that the Stage 2 audit may proceed.

• If an organization site is placed on QNBH status after a Stage 2, transfer, transition, or recertification audit, but before the certificate is issued,
• The CB shall immediately suspend the existing certificate, if applicable.
• The CB shall issue the new certificate in accordance with the Rules.
• The CB shall then immediately place the new certificate in suspension in accordance with the Rules. If applicable, the suspension of the previous certificate shall be removed. Material Management Operations Guideline/Logistics Evaluation (MMOG/LE) Organizations shall use Global MMOG/LE – version 4 to integrate evaluation of delivery performance into their quality management system. Evaluation of integration effectiveness shall be based on evidence that the organization has a process in place that includes elements such as

- Internal auditors identified.
- An established schedule for self-assessment (including evidence of schedule adherence).
- Timely submission of the completed self-assessment to FCA US.
- A defined continuous improvement process (including evidence of goal setting and performance evaluation).
- A defined corrective action process (including evidence of actions taken and verification of effectiveness).
- *Progress monitoring*: Evaluation shall be by self-assessment. The self-assessment shall be conducted annually, but may be repeated as needed. **Note**: FCA may choose to conduct a MMOG/LE audit at any time. The self-assessment shall be conducted using the "full" self-assessment spreadsheet tool from Global MMOG/LE – version 4. The results of the annual self-assessment shall be submitted to FCA US through the DRIVe system (accessible through eSupplierConnect) between May 1 and July 31 of the current calendar year. A copy of the completed spreadsheet shall be retained. Questions concerning MMOG/LE should be directed to FCA US Supplier Delivery Development at scmsdd@fcagroup.com.

9.1.3–9.2.2.1 No additional requirements.

9.2.2.2 *Quality management system audit*: The scope of the annual audit program shall include a review of a minimum of two Product Control Plans for FCA US parts, where applicable.

9.2.2.3 *Manufacturing process audit*: Layered Process Audits Organizations supplying production parts or components to FCA US shall conduct Layered Process Audits (LPA) on all elements of manufacturing and assembly lines that produce production parts or components for FCA US. These shall include both Process Control Audits (PCA) and error-proofing verification (EPV) audits. Organizations shall provide evidence of compliance to the following requirements:

- Audit process shall involve multiple levels of site management, from line supervisor up to the highest level of senior management normally present at the organization site.
- A member of site senior management shall conduct process control audits at least once per week. All members of site senior management shall conduct process control audits on a regular basis.
- Delegation of this activity will not be accepted with the exception of extenuating circumstances. **Note**: Frequent travel is an example of an extenuating circumstance. Site management personnel whose responsibilities include frequent travel may be excused from scheduled participation in layered process audits, but should participate whenever possible.
- The organization shall have a documented audit structure with auditor level and frequency of inspection.
- PCAs shall be conducted at least once per shift for build techniques and craftsmanship-related processes.

- EPV audits shall be conducted at least once per shift, preferably at the start of shift. Compliance charts shall be completed once per quarter and maintained for the life of the program. The following metrics shall be included:
  - Audit completion by all auditing layers.
  - By-item percentage conformance by area.
- Reaction plans shall be in place to immediately resolve all nonconformances. The organization shall show evidence of immediate corrective action, containment (as required), and root cause analysis (as required). A separate communication procedure is required to address reoccurring nonconformances. Specific areas of focus shall include the following:
  - Resolution of nonconformances.
  - Escalation of issue for management review.
  - Lessons learned. Layered process audits are not required for specific materials, parts, or assemblies produced on such an infrequent or irregular basis that it would prohibit establishing a regular, weekly audit schedule.
  - Such infrequently or irregularly produced materials, parts, or assemblies shall be subject, at a minimum, to a process audit at start-up and shutdown of each production run.
- Organizations shall evaluate and document the applicability of this exception for each material, part, or assembly under consideration based upon the production schedule for all customers.
- The evaluation document shall be maintained as an organization-controlled record (7.5.3.2.1), reviewed annually, and updated as required. Organizations shall use CQI-8: Layered Process Audits Guideline, 2nd Edition to establish a Layered Process Audit program. The program shall be administered under the guidance of a competent manufacturing process auditor as defined in IATF 16949 Sanctioned Interpretation no.4 for Section 7.2.3. Special Process Assessments Organizations shall evaluate the effectiveness of each of the applicable special processes listed below with the associated AIAG manual:
  - Heat Treating – CQI-9 Special Process: Heat Treat System Assessment, 3rd Edition*.
  - Plating – CQI-11 Special Process: Plating System Assessment.
  - Coating – CQI-12 Special Process: Coating System Assessment.
  - Welding – CQI-15 Special Process: Welding System Assessment.
  - Soldering – CQI-17 Special Process: Soldering System Assessment.
  - Molding – CQI-23: Special Process: Molding System Assessment.
  - Casting – CQI-27: Special Process: Casting System Assessment*.

*See "Special Process Assessments – Additional Considerations" below.

- Evaluation of implementation effectiveness shall be based on evidence that the organization has a process in place that includes elements such as
- Auditors identified.

- Schedule for self-assessment in place (including evidence of schedule adherence).
- Monitoring of progress.
- Defined corrective action process.
- Organization-controlled record keeping (7.5.3.2.1).
- Supplier development process (8.4.2.5) identified for applicable suppliers to the organization. Pursuant to IATF 16949 clauses 8.4.1.3 and 8.4.3.1 together with their associated FCA US customer-specific requirements, this requirement shall also apply to suppliers to the organization who employ the above-listed special processes. Organizations shall evaluate their manufacturing processes, and the manufacturing processes of their suppliers, to establish and document the scope of applicability of this requirement. This document is an organization-controlled record (7.5.3.2.1). Evaluation shall be by self-assessment. The self-assessment shall be conducted annually, but may be repeated as needed. The self-assessment may be conducted as part of the organization's internal quality audit or conducted separately. Assessment by a competent second-party auditor (7.2.4) will satisfy the self-assessment requirement for suppliers to the organization. *Special Process Assessments – Additional Considerations CQI-9*: Organizations shall submit a completed self-assessment to FCA US Supplier Operations on an annual basis.
- Completed assessments shall be submitted to the following SharePoint site: https://partners.chrysler.com/sites/psqcentral/CQI9/SitePages/Home.aspx.
- Submissions shall be in English
- Submissions shall be identified by the Organization name
  - Organization location o Applicable FCA US Supplier Manufacturing Location Codes (SMLCs)
  - Year of submission
- Suppliers to an organization (i.e., sub-tier suppliers) may submit completed self-assessments directly to FCA US Supplier Operations after reviewing the self-assessment with their customer. CQI-27: Organizations shall complete initial implementation of a casting self-assessment program by December 9, 2018. Self-assessment program administration is subject to the exemptions identified in Tables 7, 8, and 9 of Appendix B (they are not shown here).

9.2.2.4 *Product audit*: Continuing conformance inspection and tests shall be performed in conformance with to PF-8500 and the Global Product Assurance Testing manual during the model year to assure production items or products continue to meet specified requirements and tolerances unless waived in writing by the FCA US Release Engineer. Any such waiver shall be subject to annual review and renewal. FCA US may implement a Launch Inspection Program (LIP) project for inspection of organization supplied parts and material that Supplier Operations suspects may be at risk of nonconformance. Upon implementation of an LRM project, the organization shall cooperate with this FCA action in accordance with SQN-A0490 Launch Risk Mitigation (LRM).

9.3–9.3.2 No additional requirements.

9.3.2.1 *Management review inputs* – supplemental: Output from customer-specific requirements to the following sections shall provide management review input:

- Design and development planning – Supplemental (8.3.2.1)
- Supplier quality management system development (8.4.2.3)
- Customer satisfaction – Supplemental (9.1.2.1)
- Quality management system audit (9.2.2.2)
- Manufacturing process audit (9.2.2.3)
- Automotive warranty management (10.2.5).

9.3.3.1 No additional requirements.

Clause 10: Improvement

10.1 No additional requirements.

10.2 *Nonconformity and corrective action*: The Global Issue Management (GIM) process and system shall be used by all organizations providing parts and components to FCA US to document corrective action, unless otherwise specified by the governing FCA US business process. Application of the GIM process and system (e.g., response timing) shall conform to the governing FCA US business process.

10.2.3–10.2.4 No additional requirements

10.2.5 *Warranty management systems*: Automotive Warranty Management (AWM) Organizations providing production and non-exempt service parts and components to FCA US shall support improvement in customer satisfaction through pursuit and achievement of warranty reduction targets established by FCA US, where applicable. This shall be accomplished by active participation in the Supplier Associated Warranty Reduction Program (SAWRP). Organizations shall use CQI-14: Automotive Warranty Management, 3rd Edition to integrate warranty into their quality management system. Evaluation of integration effectiveness shall be based on evidence that the organization has a process in place that includes elements such as

- Internal auditors identified.
- An established schedule for self-assessment (including evidence of schedule adherence).
- A defined continuous improvement process (including evidence of goal setting and performance evaluation).
- A defined corrective action process (including evidence of actions taken and verification of effectiveness).
- Organization-controlled record keeping (7.5.3.2.1).
- Progress monitoring (including monthly evaluation of organization's performance to warranty reduction targets established by FCA US).
- A supplier development process (8.4.2.5) identified for applicable suppliers to the organization.

  Note: When organizations manage warranty at a corporate level, individual organization sites requiring evidence of compliance to this requirement may reference CQI-14 compliant corporate processes as they pertain to the products and processes at their sites. Evaluation shall be by self-assessment.

The self-assessment shall be conducted annually, but may be repeated as needed. The self-assessment may be conducted as part of the organization's internal quality audit or conducted separately. The self-assessment shall be conducted using the self-assessment spreadsheet tool from CQI-14. The completed spreadsheet shall serve as a record of the self-assessment. Implementation of Automotive Warranty Management shall proceed in three stages:

1. Organization identifies and implements necessary changes to quality management system processes, trains responsible personnel, and conducts initial, "baseline" self-assessment.
2. Organization establishes internal performance goals, develops prioritized corrective action plan to achieve these goals, and prepares an assessment schedule.
3. Organization monitors performance, continues with self-assessments, and updates corrective action plan as required to meet FCA US requirements and internal improvement goals or maintain goal-level performance. Implementation timing for organizations (either new suppliers or current suppliers to FCA US) is summarized in Table 3.1.

AWM Exceptions The following temporary exceptions apply to organizations that would otherwise be required to implement AWM:

1. Emergency Assumption of Business – Organizations who assume production of parts or components at FCA US's request under emergency conditions are exempt from AWM requirements for six months for these parts or components. The "New Supplier/Existing Program" requirements (Table 3.1) shall apply thereafter.
2. Financially Distressed Suppliers – Organizations that have been identified by FCA US Supplier Relations as being financially distressed may, with FCA US Supplier Operations senior management approval, suspend AWM

TABLE 3.1

Implementation Timing for Automotive Warranty Management Requirements

Organization's Relationship to FCA US	Existing Vehicle Program	New Vehicle Program
New supplier	Complete implementation through Stage 2 within six months of award of business. Implementation through Stage 3 to follow within six months of start of production	Complete implementation through Stage 2 before Commercial Launch. Implementation through Stage 3 to follow within 6 months of Commercial Launch
Current supplier	Full implementation through Stage 3 required	Follow timing for "New Supplier/ New Vehicle Program" (above) for new parts or components

actions. Such action is considered temporary and will be subject to periodic review by FCA US Supplier Operations and FCA US Supplier Relations.

3. Organizations that have been identified by FCA US Purchasing management as exempt from IATF 16949 certification are also exempt from FCA US AWM requirements. However, Mopar parts or components installed on production vehicles at an assembly plant, a Mopar Custom Shop, or a dealership at time of sale are considered "production" parts and subject to AWM requirements regardless of the organization's certification status. Implementation is not required of organizations producing modular assemblies or other products that cannot have warrantable repair assigned to their activity.

4. Implementation is not required of organizations producing parts or components in commodity groups with historically low warranty levels. A list of these low warranty commodity groups is available from the FCA US web page "Supplier Warranty Management – WIS, EWT, GCS, QNA," available in eSupplierConnect. Organizations whose volume of parts or components supplied in a specific commodity is of low significance may be exempted from FCA US AWM requirements for that commodity. The determination of exemption eligibility for a specific organization-commodity combination is the responsibility of the FCA US Supplier Quality Operations Warranty group.

Note: Questions concerning the program eligibility of individual organizations or commodity groups should be directed to the FCA US Supplier Quality Operations Warranty group at sqwarr@fcagroup.com.

10.2.6 *Customer complaints and field failure test analysis*: Returned Parts Analysis Organizations that provide production or non-exempt service parts or components shall participate in the review, testing, and analysis of returned components in accordance with PS-11346 and shall include analysis of the interaction of embedded software, if applicable. Technical Support Organizations that provide production and non-exempt service parts and components shall provide all necessary support to FCA US in the investigation and resolution of supplier-associated warranty issues.

10.3–10.3.1 No additional requirements.

Appendix A identifies the list of the Bulk Metallic Commodities in Table 5 as

Code	Name
03AB	Hot Rolled Steel
03BA	Cold Rolled Steel
03CC	Galvanized Steel-Both Sides
03IA	Steel – Special Shapes
03KF	Welded Carbon Steel Tube
03NA	Hot Rolled Carbon-Bars
03RA	Welding Wire, Rods
05AD	Flat Rolled Aluminum
05AG	Aluminum Braze Sheet

Appendices B and C identify the CSR section exemptions for the bulk metallic commodities in Table 6 as

IATF 16949 Section	FCA US Customer-Specific Requirement
8.2.3.1.2 Customer-designated special characteristics	The Shield; also, The Diamond
8.3.5.2 Manufacturing process design input	PFMEAs and Control Plans
8.3.4.2 Design and development validation	DV; PV
8.3.4.4 Product approval process	Process Approval; PPAP
8.6.2 Layout inspection and functional testing	Annual Layout
8.6.3 Appearance items	Appearance Master Samples
9.2.2.3 Manufacturing process audit	Layered Process Audits
10.2.5 Warranty management systems	Automotive Warranty Management (AWM); AWM Exceptions
10.2.6 Customer complaints and field failure test analysis	Returned Parts Analysis
APPENDIX B: EXCEPTIONS TO CQI-27	APPENDIX C: CHANGE HISTORY

GENERAL MOTORS COMPANY (GM)

GM requires their suppliers to meet additional requirements from the base of standards (ISO 9001 and IATF 16949). These requirements fall into two categories. They are (a) Built-In Quality Supply (BIQS) System and (b) detailed specific requirements on a per element basis.

GM's BIQS System

Just like many companies, GM has its own prequalifying requirement of doing business with any one supplier. That is, a system that must continually look for ways to increase value and reduce waste throughout its supply chain. Like many manufacturers, GM is under pressure to keep ahead of the increasing challenge to meet demands for ever better, faster, and cheaper products. That system for the GM is the BIQS with 29 robust elements replacing the old Quality Systems Basics (QSB) with its 11 quality elements.

BIQS is the result of unprecedented recall problems that have plagued GM. In 2015, Isidore (2015) reported that these problems cost GM over 4.1 billion dollars. The BIQS is the result of a strategic plan to improve quality throughout the company by incorporating input(s) from their supply chain (Isidore, 2015).

As a system, BIQS assesses suppliers according to 30 different elements. That's in addition to IATF 16949 certification and metrics like quality problem reporting and resolution (PRR), field actions, disruptions, and severity score. Suppliers must achieve a Level III score or higher to be a certified supplier, with each element scored based on a green, yellow, or red rating. Among those 30, the high-risk items must be checked during each shift, and they are

- Critical operations
- Customer complaints related issues

- Problem-solving
- Fast response events.

So, what specifically do GM auditors expect to see in the quality focused checks when they visit a facility? The most important thing is that the items above are checked every single shift. To facilitate this process, the auditor may use some kind of a checklist or some other convenient tool. In addition, these checks can be part of a layered process audit (LPA) or conducted separately. This preeminence of checking is more than a typical random sampling. Rather, it is a priority of any LPA. So, there is no confusion as to what to do, when it is recommended that the LPA may include a sampling of rotated and randomized questions. In other words, the audit should also include a set of fixed questions focused on critical controls.

The criteria for success (green rating) are based on the following:

1. Your system clearly shows you are conducting them.
2. People are aware that they have to do the checks.
3. You have a robust tool implemented to track and ensure compliance.
4. Your system is working correctly.

Furthermore, the intent of the BIQS is to bring much needed awareness to GM's suppliers about their quality expectations. In doing so, it brings forth best practices that a good quality management system can implement in any situation to make "lasting" improvements in quality. Therefore, the BIQS focuses on three major strategies. They are

1. *Utilization of internal audits*: Specifically, the utilization of the layered audit (LA). Focus of the audit is to identify the frequency, schedule, findings, and corrective actions – if needed. Furthermore, the power of the LA process audit is its simplicity and that it brings "power" to both management and floor employees to bring about issues, concerns, and problems in such a way that they are *fixed right the first time*. This is accomplished at least by
   a. Assessing the supplier's compliance to standardized processes
   b. Assigning management the responsibility for assuring the effective implementation and adherence to scheduled audits
   c. Identifying opportunities for continuous improvement
   d. Providing coaching opportunities
   e. Requiring management to actively participate in the audit process on the shop floor on a frequent basis
   f. Including customer-specific and quality-focused checks reviewed by all layers including management
   g. Assigning management the responsibility of ensuring that effective corrective actions and counter measures are in place.
2. *Find, define, and analyze the risk for the applicable operations, using the FMEA (AIAG or the VDA* format) *methodology as a minimum requirement.* It is very important here to emphasize that the PFMEA must

be completed by a team of stakeholders (cross-functional team) and not strictly by engineers. So, every FMEA must have its own team and not a generic team to do all FMEAs. This is very important to keep in mind because this approach must also be consistent throughout all the FMEAs in defining and analyzing the severity, occurrence, and detection ranking tables for prioritization of risk.

3. *Implement a fast response* with an appropriate and applicable problem-solving methodology (i.e., 5-Why's, 3 × 5 Why's, G8D, or some other formal methodology). This response must be monitored and reported as necessary with appropriate evidence. The triggered action is based on high severity of a nonconformance, test failure safety incident, customer, or any other problem.

It is imperative to recognize here that fast response requires verification (authorization by a GM representative, before the action is initiated) that suspect parts are contained, a disciplined root cause investigation is conducted, short- and long-term solutions are considered, and updates are made to document process instructions and controls. Validation is also required (approved by a GM representative) before the action is approved for the "fix" of the problem and implemented as complete and accepted.

BIQS is being phased out by the end of 2020 end no later than the spring of 2021. Its replacement is being discussed as of this writing to have included minor changes of the current BIQS with an addition of the manufacturing site assessment. The pending title is *1927 Global Supplier Quality*. Because it is not in the final stage yet, we have chosen not to address any of its requirements. However, a good update source is http://girekotu.blog.fc2.com/blog-entry-2083.html.

GM's Primary Requirements

To supplement the ISO 9001, IATF 16949, and the BIQS requirements, GM has additional conditions and clarifications to do business with its suppliers. The structure of these qualifications follows the ISO 9001. The information on this section is based on https://www.iatfglobaloversight.org/wp/wp-content/uploads/2019/06/IATF-16949-GM-CSR-May-2019-_V5.pdf GM. Retrieved on January 7, 2020. They are

4 Context of the organization

4.1–4.4.2 No additional requirements.

5 Leadership

5.1–5.3.2 No additional requirements.

6 Planning

6.1–6.3 No additional requirements.

7 Support

7.1–7.5.3.2 No additional requirements.

7.5.3.2.1 *Record retention*: The organization's business records shall be retained as specified in GMW15920. Organizations can purchase GMW documents from IHS at www.global.ihs.com

7.5.3.2.2 No additional requirements.

8 Operation

8.1–8.2.3.1.1 No additional requirements.

8.2.3.1.2 *Customer-designated special characteristics*: The organization shall follow General Motors Key Characteristic Designation System Process GMW15049. Key Characteristics shall be applied as per IATF 16949:2016 8.3.3.3 Special Characteristics.

8.2.3.1.3–8.3.3 No additional requirements

8.3.3.1 *Product design input*: All operations shall be analyzed for risk using a PFMEA. Product requirements shall be identified, and failure modes comprehended in the PFMEA. Risk Priority Number (RPN) values shall be consistently applied using Severity, Occurrence, and Detection ranking tables. Severity shall be based on all risks such as organization risk, customer risk, and end-user risk.

8.3.3.2 No additional requirements

8.3.3.3 *Special characteristics*: The organization shall have a process to identify critical operations within their manufacturing process.

8.3.4–8.3.4.3 No additional requirements

8.3.4.4 *Product approval process*: The organization shall comply with the AIAG Production Part Approval Process (PPAP) manual and GM 1927-03 Quality SOR to meet this requirement.

8.3.5–8.3.5.1 No additional requirements.

8.3.5.2 *Manufacturing process design output*: The organization shall have a method to identify, control, and monitor the high-risk items on those critical operations. There shall be rapid feedback and feed forward between inspection stations and manufacturing, between departments, and between shifts.

8.3.6 No additional requirements.

8.3.6.1 *Design and development changes – supplemental*: All design changes, including those proposed by the organization, shall have written approval by the authorized customer representative, or a waiver of such approval, prior to production implementation. See also AIAG Production Part Approval Process (PPAP) manual.

8.4–8.4.2.4 No additional requirements

8.4.2.4.1 *Second-party audits*: Second-party auditors performing QMS audits must meet the requirements in clause 7.2.4 Second Party Auditor Compliance in IATF 16949:2016 plus meet these additional requirements: (a) The organization must be IATF 16949:2016 certified and not on suspension, and (b) the second-party auditor must be a qualified ISO Lead Auditor, or a qualified internal auditor with evidence of their successful completion of training, and a minimum of five internal ISO/TS 16949:2009 and/or IATF 16949:2016 audits under the supervision of a qualified lead auditor. The organization may conduct (second-party) audits of their supplier per their supplier development risk management analysis.

For initial certifications, the first second-party audit should use the initial audit days from Table 5.2*. For subsequent second-party audits, use the recertification days Table 5.2*.

*See Automotive Certification Scheme for IATF 16949, Rules for Achieving and Maintaining IATF Recognition, Section 5.2, Table 5.2 Minimum audit days.

The second-party audits shall identify an acceptable passing level and include a scoring or ranking to determine which suppliers have passed. The organization shall have documented evidence that they review and follow up on all nonconformances identified in the second-party audit with the intent to close these nonconformances.

8.4.2.5 *Supplier development*: When a supplier to an organization is so small as to not have adequate resources to develop a system according to IATF 16949:2016 or ISO 9001:2015, certain specified elements may be waived by the organization. The organization shall have decision criteria for determining "specially designated small suppliers." Such decision criteria shall be in writing and applied consistently in the application of this provision. The existence and use of such decision criteria shall be verified by third-party auditors.

Note 1: ISO 9001:2015 and IATF 16949:2016 MAQMSR contain fundamental quality management system requirements of value to any size of provider of production materials, production, service, and accessory parts, or heat treating, plating, painting, or other finishing services. There are a number of methods to implement a compliant system, so it is recognized that a simpler quality management system approach could be used for the smaller suppliers of organizations to which IATF 16949:2016 clause 8.4.2.3 applies.

Note 2: "Small" may also refer to volume supplied to automotive.

8.4.3–8.5.1 No additional requirements

8.5.1.1 *Control plan*: General Motors does not provide waivers to organizations for control plan approval because General Motors signatures on the control plan are not required. The organization shall provide measurement, test, and inspection data which demonstrates that control plan requirements, sample sizes, and frequencies are being met when requested. Sample sizes and frequencies shall be determined based on risk and occurrence of failure modes, and to ensure that the customer is adequately protected from receiving the product represented by the inspection/tests before the results of the inspection/tests are known.

8.5.1.2 *Standardized work – operator instructions and visual standards*: Standardized work should include the what, how, and why tasks are performed. All standardized work shall be followed. Visual standards throughout the facility shall be common, including between facilities building the same platform/product for global quality. Visual standards shall be clearly communicated to all team members that are affected and referenced in the standardized work. Visual standards that differentiate "good" from "bad" shall satisfy customer requirements and be controlled.

8.5.1.3–8.5.1.5 No additional requirements.

8.5.1.6 *Management of production tooling and manufacturing, test, inspection tooling, and equipment*: Where warehouses or distribution centers (distributors) are remote sites, the requirements for management of production tooling may not be applicable.

8.5.1.7–8.5.6 No additional requirements.

8.5.6.1 *Control of changes – supplemental*: The documented process shall require consideration of a production trial run for every product and process change. Results of the trial run shall be documented.

8.5.6.1.1 *Temporary change of process controls*: The organization shall have a process for both bypass and deviation. The alternative actions identified on the bypass list shall be customer approved and shall be reviewed using the methodology of the PFMEA to identify the risk. This review shall be documented.

8.6–8.6.1 No additional requirements.

8.6.2 *Layout inspection and functional testing*: Unless specified otherwise by a GM Procuring Division, there is no customer established frequency for layout inspection after receiving production part approval (PPAP).

8.6.3–8.7.2 No additional requirements.

9 Performance evaluation

9.1–9.1.1 No additional requirements.

9.1.1.1 *Monitoring and measurement of manufacturing processes:* The organization shall have a method for the employee to call or notify for help when an abnormal condition on the equipment or product occurs. A method to call or notify shall be available in all operational areas of the organization. Sufficient alarm limits shall be established for escalation of abnormal conditions and shall match the reaction plan identified in the product's control plan.

9.1.1.2–9.1.2 No additional requirements.

9.1.2.1 *Customer satisfaction – supplemental*: New Business Hold. The organization shall notify their CB within 5 business days of receiving notice of special status condition of GM New Business Hold – Quality. The CB shall take the decision to place the organization on immediate suspension

*Upon receiving notice of GM New Business Hold – Quality (NBH).

*See Automotive Certification Scheme for IATF 16949, Rules for Achieving and Maintaining IATF Recognition, Section 8.3.

4. In the event of certification suspension as a result of an organization receiving notice of General Motors New Business Hold – Quality, the

organization shall complete a corrective action plan. The organization shall submit the corrective action plan to the CB and to the affected customer(s) within ten business days of the effective date of the NBH. The corrective action plan of the organization shall be consistent with the affected customer requirements including correction steps, responsibilities, timing information, and key metrics to identify effectiveness of the action plan.

5. Before any suspension can be lifted, the CB shall take the decision to conduct an on-site special audit of appropriate length to verify effective implementation of all corrective actions. The special audit must be conducted within 90 calendar days from the notice of New Business Hold – Quality.

   If suspension is not lifted within the maximum of 110 calendar days from the notice of New Business Hold – Quality, the CB shall withdraw the IATF 16949 certificate of the organization. Exceptions to this withdrawal shall be justified in writing by the CB based upon its on-site review of the effectiveness of the organization's corrective action plan and agreement obtained in writing from the authorized GM customer representative.

Note 1: When an organization is placed in NBH after a recertification (or initial) site audit but before the certificate is issued,
- The CB shall issue the certificate in accordance with the IATF Rules.
- The CB shall then place the new certificate in immediate suspension with the rules for lifting such suspension appropriately applied.

BIQS Requirements Organizations shall achieve and maintain BIQS Level of 3, 4, or 5. The organization whose BIQS Level falls below Level 3 shall notify its CB within 5 business days after falling below the stated requirement. If the organization fails to notify their CB, the CB shall issue a minor nonconformance against IATF 16949:2016, clause 9.1.2.1. [Warning: these requirements may change due to the new standard which will take effect no later than the spring of 2021.]

The CB shall issue a major nonconformance against IATF 16949:2016, clause 9.1.2.1, when they are notified (or discover), the organization is at a BIQS Level 1 or 2. The CB shall conduct an on-site special audit. Organizations that have not had their initial IATF 16949 certification and are BIQS Level 1 or 2 shall not be issued a nonconformance.

To close this major nonconformance during the on-site special audit, the organization shall have either (a) achieved BIQS metrics of Level 3, 4, or 5; or (b) a documented action plan, confirmed by the GM SQE or SQE designee, detailing the steps, improvements, with target dates, being made to achieve BIQS Level of 3, 4, or 5.

Note: The GM system source ability report will indicate a BIQS Level of 1 or 2 for those organizations not meeting the BIQS requirements. The organization shall notify its CB within 5 business days after being placed in Controlled Shipping – Level 2 (CS II) Status. The CB is not required to issue a nonconformance for an organization placed in CSII status.

For CSII activities that are open during an audit, the organization's CB shall verify that an effective corrective action is in process and, if closed, that the corrective actions have been implemented and read across to the entire organization's site for similar processes and/or products. The organization's CB shall also investigate any CSII activities that have occurred and were closed between surveillance audits.

Note: The GM condition of CS II (Controlled Shipping – Level 2) is a performance indicator of problems in an organization's product realization process. The CSII condition should have resolution, or credible resolution and corrective plans in place, which are confirmed by the customer.

9.1.3–9.2.2.2 No additional requirements

9.2.2.3 *Manufacturing process audit:* The organization shall incorporate an internal layered process audit process to assess compliance to standardized processes, to identify opportunities for continuous improvement, and to provide coaching opportunities. The layered process audit is led by management who are competent to conduct the audits. The process shall include

1. A schedule including frequency of audits and locations of planned audits.
2. Audit layers must be used and include different levels of employees, including top management.
3. Customer complaints or rejections trigger a layered audit on the process that was cause of the issue.
4. All departments within the organization.
5. All findings are recorded and measured for improvement.
6. Findings that cannot be corrected during the audit shall move to an action plan for monitoring to closure.
7. Records of audits shall be maintained.
8. Layered audit questions shall be reviewed periodically and changed if needed to focus on the organization's weaknesses.
9. Layered process audit shall be done as part of corrective action verification activities.
10. In addition to layered process audits, the organization shall audit specific manufacturing processes (see Note 2) annually to determine their effectiveness. Applicability and effectiveness of these processes shall be determined utilizing the most current version CQI standard (see the chart below). The effectiveness evaluation shall include the organization's self-assessment, actions taken, and that records are maintained.

Note 1: The assessment must be performed by a competent auditor. An auditor is competent if they meet the following requirements:

- They shall be a qualified ISO 9001:2015 Lead Auditor, or a qualified internal auditor with evidence of their successful completion of training, and a minimum of five internal ISO/TS 16949:2009 and/or IATF 16949:2016 audits under the supervision of a qualified lead auditor.

- They shall have a minimum of 5 years' experience working with the process that is being audited or a combination of experience and education in the specific process.

Note 2: Audit findings must be addressed in an action plan, with champion(s) assigned and reasonable closure dates.

CQI Standards:

- Heat Treating Processes CQI-9 Heat Treat System Assessment
- Plating Processes CQI-11 Plating System Assessment
- Coating Processes CQI-12 Coating System Assessment
- Welding Process CQI-15 Weld System Assessment
- Plastics Molding Processes CQI-23 Molding System Assessment
- Solder Processes CQI-17 Soldering System Assessment
- Casting Process CQI-27 Casting System Assessment.

9.2.2.4 *Product audit*: The organization shall perform quality focused checks on each shift.

The organization shall have a process for final inspection and/or Customer Acceptance Review & Evaluation (CARE). GP-12 shall be performed as required during launch and until released by the organization's assigned SQE or designate and per GM 1927–28 Early Production Containment (GP-12).

1. Final inspection shall be performed on all finished product prior to shipping. This inspection can be 100% inspection or less, based on risk.
2. GP-12 inspection checks shall be included at an upstream inspection station (final inspection/CARE).
3. Quality checks shall be included in standardized work. Point, touch, listen, and count inspection methods are incorporated.
4. Successive production/quality checks shall be increased in cases of high risks such as model launch, pass through components and characteristics pass through, major changes, shutdown (see clause 8.5.1.4), or customer feedback.

9.3–9.3.3.1 No additional requirements

10 Improvement

10.1–10.2.2 No additional requirements

10.2.3 *Problem-solving*: The organization's documented problem-solving process shall include

1. Tracking of issues through closure.
2. Daily review of issues by a multi-disciplined team including plant management.
3. Daily reviews are documented.
4. All levels of the organization are included in the problem-solving process.
5. Robust method to identify the verifiable root cause(s) of each issue.

6. Timely closure of corrective action(s) including exit criteria.
7. Initial containment is well documented using a containment worksheet or similar

10.2.4 *Error-proofing*: Error-proofing devices shall be tested to failure or simulated failure at the beginning of each shift at a minimum, otherwise according to the control plan. In the event of error-proofing device failure, a reaction plan that includes containment should be included in the control plan. The organization shall keep a list of all error-proofing devices and identify which can be bypassed and which cannot (also see clause 8.5.6.1.1). The bypass determination shall consider safety, severity, and overall RPN rating.

10.2.5–10.3 No additional requirements

10.3.1 *Continual improvement – supplemental*: The organization shall have a process for effective review of PFMEA of all manufacturing parts and processes to occur annually at a minimum. This review shall consider, at a minimum, critical, safety, and high-risk items. The organization shall incorporate tools such as reverse PFMEA or other similar methods to assist in the PFMEA review. PFMEA review output shall include an updated PFMEA, record of the changes made (or record that no changes were made), and identification of the team involved in the review.

Critical, safety, and high-risk items (such as priority from Risk-Limiting Method, high RPN, or equivalent) shall have an action plan which includes recommended actions, responsibility, and timing.

Reviewing a PFMEA for corrective action process does not meet the requirement of annual review unless there is evidence that critical, safety, and high-risk items are considered in addition to the corrective action issue. A proactive review approach is required.

FORD'S CUSTOMER-SPECIFIC REQUIREMENTS

Ford requires their suppliers to meet additional requirements from the base of standards (ISO 9001 and IATF 16949). These requirements fall into three categories: (a) manufacturing site assessment, (b) Q1 3rd ed. and (c) detailed specific requirements on a per element basis.

Manufacturing Site Assessment (MSA)

Ford requires at least once a year (more often, if necessary) to have a complete assessment of a supplier's facilities. The assessment is conducted by the supplier and reviewed by a Ford representative – usually the STA (Supplier Technical Assistance). Ford has developed their own questionnaire with both requirements and expectations. It may be requested from Ford and/or downloaded from the supplier portal in the covisint site. The major areas of the survey are as follows:

1. Planning for manufacturing and process capability

2. Sub-supplier quality management.

It is expected by the end of 2020 that the requirements of the MSA survey will be loosen up. The frequency will be depended on special exceptions such as stop shipment or specific supplier problems. No longer the yearly survey will be necessary.

Q1 3RD ED.

Having a Ford Q1 3rd edition certification suggests that a supplier's facility or manufacturing site has achieved excellence in four important areas. They are quality performance (30 points), capable systems (30 points), warranty performance (20 points), and delivery performance (20 points).

The focus of these four areas is to standardize and make it easier for suppliers to demonstrate objective evidence for their quality. The Q1 3rd ed. has its own evaluation checklist and numerical value of 0–100 total points. The minimum passing is 80 points, however, that may be adjusted to reflect specific situations of a particular supplier. In addition to this minimum, a supplier may get "extra" designation (recognition) in the evaluation process, if the performance is consistent for a period of six (6) months and has a cumulative score thusly:

- Q1 Gold >95 points
- Q1 Silver 90–95 points
- Q1 Basic 80–90. If the supplier has <80 points, they are NOT Q1 certified. Q1 score must be at least 80 points to maintain Q1, with active metrics in all categories. A Q1 Supplier Site with a score below 80 points or carryover metrics with zero score will be recommended for Q1 Revocation

It is important to recognize that the Q1 scoring is updated via the Supplier Improvement Metrics (SIM) refresh weekly. Q1 scoring is not updated between the weekly refresh, even though some metrics may be updated during the week. The objective measures are as follows:

- Quality performance
  - Parts per million (PPM) commodity performance (production and or service)
  - Quality rejects (QRs) trend (production and or service)
  - Stop ship(s).
- Capable systems
  - Q1MSA (Q1 Manufacturing Site Assessment). Typical measurement items are
    - Implementation of fundamental quality management system processes
    - Change management
    - Corrective action
    - Failure mode avoidance

- Sub-tier supplier management
- Manufacturing feasibility.

A Q1MSA is not a onetime event, but an enabler for continual improvement. It is to be conducted at least once a year by the supplier and reviewed by the Supplier Technical Assistant (STA).

- Industry Standards
  - Third-Party Certification
    - IATF 16949 or ISO 9001 Quality Management System
    - ISO 14001 Environmental Management System.
- Self-Assessment
  - Global Materials Management Operations Guide/Logistics Evaluation (MMOG/LE Level A required) (Annual Self-Assessment)
- APQP Launch Performance
- Warranty Performance
  - 0 Field Service Actions in most recent 6 months with any supplier responsibility
  - R/1000 (Repairs per 1000 Vehicles) score of 3 or higher
- Delivery Performance
  - For all delivery ratings considered for Q1, the 6-month weighted average AND the last three available scored months must all be 81 or greater (which may include the immature month).

Scoring

- Quality Performance – 30 Q1 points
  - Production commodity (ppm) 10 points
  - Service commodity (ppm) 10 points
  - Production QR rate 10 points
  - Service QR rate 10 points
  - Stop ship 20 points

$$\text{Normalizing equation} = \frac{\text{Sum of earned metric points}}{\text{Sum of possible metric points}} \times \text{Category Q1 points}$$

Example of calculating the individual performance category of each metric.

$$\frac{\text{Total metric points earned}}{\text{Total metric points possible}} \times 30 = \text{Points recorded in the specific category}$$

Inactive metrics are excluded from both the numerator and denominator of the calculation above. If there are no active metrics in a category, the category will be scored at 80% of the possible Q1 point value. Very important is Q1 *is not* attainable without some active metrics in every category. The temporary 80% score allows other categories to place the total Q1 score above or below 80 Q1 points. The function of the auditor is to make sure that the certification to the Q1 standard is current and has identified action plan(s) for any nonconformance items.

- Capable systems 30 Q1 points
  - Q1 manufacturing site assessment 15 points
  - Industry standards 10 points
  - APQP launch performance 5 points.
- Warranty performance 20 points
  - Field service actions 10 points
  - R/1000 performance 5 points
  - Shared site calculation 5 points. Of the 20 points, 10 are for no FSA (Field Service Actions), 5 is for warranty based on R/1000 for 3, 6, and 9 MIS. Then, the system takes the total points for warranty divided by 15 (max total points for warranty) times 20. For example, if warranty scoring is 14, then $14/15 = 0.933$, times $20 = 18.66$ and the system would round up to 19.
- Delivery 20 points
  - Production performance 10 points
  - Service performance 10 points.

For each individual category, Ford provides a detailed explanation to accumulate the total points. We are not addressing the detailed calculations since the auditor is not responsible for the numerical values. They are interested whether or not the customer-specific requirements have been taken into account. The normalized equation shown above is used for all categories, but for the individual points per item, they are described in the Ford documentation. For more information, see SUPLCOMM@ford.com.

Specific Requirements on a per Element Basis

To have access to all the mentioned internet sites relating to Ford, one must have access to the Ford portal for the suppliers in the covisint system. The information on this section is based on: https://www.iatfglobaloversight.org/wp/wp-content/uploads/2016/12/Ford-IATF-CSR-for-IATF-16949-1May2017.pdf. Retrieved on January 7, 2020.

Clause 4: Context of the Organization

4.1–4.2 No additional requirements.

4.3 *Determining the scope of the quality management system.* The structure of this document aligns with the requirements with the applicable sections of IATF 16949. Several section headers are followed by the statement "No Ford Customer-Specific Requirement for this section" to verify that there is no auditable Ford-specific requirement for this section. The presence of this statement does not mean that no other commercial or technical requirements exist for the subject addressed in the section, or that this statement supersedes existing commercial or technical requirements.

- Tooling and equipment suppliers to Ford Motor Company are not eligible for certification to IATF 16949. Registration to ISO 9001 is acceptable.

- *Third-party registration*: To achieve Q1 (refer to https://web.qpr.ford.com/sta/Q1.html), production and service part organizations supplying product to Ford shall be third-party registered to IATF16949 through an IATF-recognized CB. The official list of IATF-recognized Certification Bodies is available through http://www.iatfglobaloversight.org/certBodies.aspx.
- A sub-tier supplier hired by the organization to perform services not directly related to a Ford Motor Company contract (e.g., floor cleaning or grass cutting) is not impacted in any way by the sub-tier supplier development or other sub-tier supplier requirements stated in IATF 16949.
- Evidence of IATF 16949 Certification Verification: Organizations shall record evidence of their certification to IATF 16949 in GSDB online available through Ford Supplier Portal https://web.gsdb2.ford.com/GSDBeans/servlet/gsdbeans.web.lib.GSDB.
- Notification of IATF 16949 Registration Status Change: Organizations shall notify Ford of any change in their IATF 16949 registration status via updating their certification information in GSDB online. Such changes include but are not limited to
  - Initial certification
  - Recertification
  - Transfer of certification to a new CB
  - Certificate withdrawal
  - Certificate cancellation without replacement.
- IATF 16949 Certification Waiver: Ford may, at its option, fully waive certain organizations from IATF 16949 certification. This waiver generally applies to those organizations whose quality management system is acceptable without certification to IATF 16949, but Ford still requires the suppliers to report any changes.
- Identification of candidate organizations for waiver from IATF 16949 certification is the responsibility of Ford. Verification and maintenance of waiver status is the responsibility of Ford.

4.3.1–4.4.2 No Ford customer-specific requirement for this section.

Clause 5: Leadership

5.1–5.1.1 No additional requirements

5.1.1.1 *Corporate responsibility*: The organization shall comply with Basic Working Conditions in the Global Terms and Conditions and the related Supplier Social Responsibility and Anti-Corruption Requirements Web-Guide https://web.fsp.ford.com/gtc/docs/hrandwc.pdf. The organization is also encouraged to adopt and enforce a similar code with Ford's Policy Letter #24 available through http://sustainability.ford.com (go to "downloads" at the bottom of the page, and then, search for "Code of Basic Working Conditions" in the list of downloads).

5.1.1.2–5.1.1.3 No additional requirements

5.1.2 *Customer focus*: The organization shall demonstrate enhanced customer satisfaction by meeting the continuous improvement requirements of Q1 3rd ed., as demonstrated in the organization's QOS (Quality Operating System).

5.2–5.3 No Ford customer-specific requirement for this section.

5.3.1 *Organizational roles, responsibilities, and authorities – supplemental*: The organization shall notify Ford Motor Company Supplier Technical Assistance in writing within ten working days of any changes to senior management responsible for product quality or company ownership.

5.3.2 No additional requirements

Clause 6: Planning

6.1–6.1.2.2 No additional requirements

6.1.2.3 *Contingency plans*: The organization shall notify the Ford receiving plant, the buyer, and the STA engineer within 24 hours of organization production interruption. The organization shall communicate the nature of the problem to Ford and take immediate actions to assure supply of product to Ford.

Note: Production interruption is defined as an inability to meet the Ford-specified production capacity volume.

Supply chain risk analysis: The organization shall have a documented Supply Risk Management Operating System in place. The organization shall ensure that its QOS Supply Risk Management process includes

- The application of the requirements for risk analysis, preventive actions, and contingency planning described in Section 6.1.2.1–6.1.2.3 of IATF 16949:2016 through the organization's supply chain.
- Documentation of the organization's supply chain (supplier name, location, parts) for all Ford-specified parts and associated raw materials
- A system to assess and monitor supply chain financial and operational risks. A list of Ford's endorsed supply chain monitoring services and tools is available through Appendix A https://web.qpr.ford.com/sta/Ford_IATF_16949_CSR_Appendix_A.pdf to assist in the establishment of the organization's QOS Supply Risk Management process
- Ford reserves the right to review the documented information of the supply chain risk assessment reviews

6.2–6.3 See ISO 9001:2015 requirements.
No additional requirements.

Clause 7: Support

7.1–7.1.3 See ISO 9001:2015 requirements.
No additional requirements.

7.1.3.1 *Plant, facility, and equipment planning capacity reporting*: Whenever the organization reports Purchased Part Capacity (Average Purchased Part Capacity – APPC, or Maximum Purchased Part Capacity – MPPC) to Ford in demonstration of compliance to the average production weekly and/or maximum production weekly (APW/MPW) capacity requirements, the organization shall use the Capacity Analysis Report to determine the values of APPC and MPPC reported.

Where equipment is not dedicated to the Ford part being reported for Purchased Part Capacity (PPC), the organization shall use either the Shared Loading Plan in the

Capacity Analysis Report or the detailed shared loading tool. The Capacity Analysis Report is available through: https://web.qpr.ford.com/sta/Capacity_Analysis_Report. xlsx. Reporting of PPC to Ford may include the following:

- Quarterly Reporting of PPC to Ford's Capacity Planning systems
- Responding to a request for quote
- Responding to a capacity study
- Capacity verification associated with PPAP
- Any other Ford request for reporting PPC.

Note: For the APPC and MPPC to be acceptable, the APPC and MPPC must meet or exceed the required capacity – APW in a 5-day operating pattern and MPW in 6-day operating pattern, respectively. Organization personnel completing the Capacity Analysis Report (CAR) are required to have completed the latest Capacity Analysis training available via https://www.lean.ford.com/cqdc/supplier_training. asp. Capacity Planners are to review the Capacity Analysis training annually. If the Capacity Analysis training is updated, Capacity Planners are required retake the Capacity Planning training and to re-register in the capacity Supplier Directory https://web.supplierdirectory.ford.com/sd/homePage.

7.1.4–7.1.5.1 See ISO 9001:2015 requirements.

No additional requirements.

7.1.5.1.1 *Measurement system analysis gauging requirements*: All gauges used for checking Ford components/parts per the control plan shall have a gauge R&R performed in accordance with the appropriate methods described by the latest AIAG Measurement Systems Analysis Manual (MSA) to determine measurement system variability. The Gauge R&R is to be completed using Ford parts. The control plan identifies which gauges are used for each measurement. Any measurement equipment not meeting the MSA guidelines must be approved by STA.

- *Family of gauges*: Where multiple gauges of the same make, model, size, and method of use and application (including range of use) are implemented for the same part, use of a single gauge R&R covering those multiple gauges (family of gauges) requires STA approval.
- *Parts and operators for gauge R&R studies*. At a minimum, (a) variable gauge studies should utilize a minimum of 10 parts, 3 operators, and 3 trials; and (b) attribute gauge studies should utilize a minimum of 50 parts, 3 operators, and 3 trials. For more information, see the Ford PPAP customer specifics for details on attribute gauge measurement systems analysis requirements (https://web.qpr.ford.com/sta/Ford_Specifics_for_PPAP.pdf)

7.1.5.2–7.1.5.3.1 See ISO 9001:2015 requirements.

No additional requirements.

7.1.5.3.2 *External laboratory*: The organization shall approve commercial/independent laboratory facilities prior to use. The acceptance criteria should be based on the latest ISO/IEC 17025 (available through ISO http://www.iso.org/), or national equivalent, and shall be documented. Accreditation to ISO/IEC 17025 or national equivalent is not required.

7.1.6–7.2 See ISO 9001:2015 requirements.

7.2.1 *Competence – supplemental*: Training shall include the appropriate Ford systems: Ford training opportunities are available through Ford Supplier Learning Institute at https://fsp.covisint.com log into Ford Supplier Portal, and then, go to the Ford Supplier Learning Institute (FSLI) application. Additional training is available through https://www.lean.ford.com/cqdc/.

7.2.2–7.5.1.1 See ISO 9001:2015 requirements.

No additional requirements.

7.5.2 *Creating and updating* **Note 1**: Where the organization uses Ford documents/instructions or other documents of external origin, the organization ensures that the appropriate revision level is used – this is either the most current version available from FSP (Ford Supplier Portal https://fsp.covisint.com) or as specified by Ford Motor Company.

Note 2: Engineering Standards may be obtained from the following sources:

IHS Markit; http://www.ihs.com/; ILI Infodisk, Inc.; http://www.ili-info.com/. If any standards are not available through the above sources, organizations should contact Ford Engineering, or for organizations with Ford Intranet access, http://www.rlis.ford.com/cgi-bin/standards/iliaccess.pl/ may provide a more complete inventory.

- Ford Engineering Specifications may be available in Ford's CAD database, TeamCenter; contact the Ford PD engineer for details. Additionally, Ford Engineering CAD and Drafting Standards (FECDS) are available through https://team.extsp.ford.com/sites/C3PNGMethods/C3PNGMethods.html.
- *Engineering Specifications (ES)*: Ford requires all manufacturing sites to report all materials per WSS-M99P9999A1, as noted in PPAP, Ford specific instructions. These requirements are detailed on Ford Supplier Portal https://fsp.covisint.com (Important Documents – RSMS Communication Package).
- *Engineering Specification (ES)*: Test Performance Requirements: The goal of ES testing is to confirm that the parts meet design intent. ES test failure shall be cause for the organization to stop production shipments immediately and take containment actions. The organization shall immediately notify Ford Engineering, STA and the using Ford Motor Company facility of any test failure, suspension of shipments, and identification of any suspect lots shipped. After the root cause(s) of ES test failure are determined, corrected, and verified, the organization may resume shipments. The organization shall prevent shipment of suspect product without sorting or reworking, to eliminate the nonconformance. These ES requirements apply equally to sub-tier suppliers.

7.5.3–7.5.3.1 and 7.5.3.2 See ISO 9001:2015 requirements.

No additional requirements.

7.5.3.2.1 Record retention

Inspection and measurement records: The organization shall retain records of process control data, product inspection data, and records of appropriate reaction actions to readings outside the specification in a recoverable format for a minimum of 2 years, available to Ford Motor Company upon request. The organization shall record the actual values of process parameters and product test results (variable or attribute). Simple pass/fail records of inspection are not acceptable for variable measurements.

- *Audits*: The organization shall retain records of internal quality system audits and management review for 3 years.
- *APQP*: The organization shall maintain the final External Supplier APQP/PPAP Readiness Assessment (Schedule A) for the life of the part (production and service) plus 1 year as part of the PPAP record.
- *Training*: The organization shall retain records of training for 3 years from the date of the training.
- *Job set-up*: The organization shall retain records of job set-up verifications for 1 year.
- Retention periods longer than those specified above may be specified by an organization in its procedures.
- *Maintenance*: The organization shall retain records of maintenance for 1 year. The organization shall retain records of measurement equipment calibration for one calendar year or superseded, whichever is longer. Ford reserves the right to modify specific record retention requirements. These requirements do not supersede any regulatory requirements.

Clause 8: Operation

8.1 *Operational planning and control statement of work*: Appropriate to the organization's responsibilities, the organization shall meet the requirements of the Statement of Work(s). There may be an Engineering Statement of Work (available from the Ford Product Development Engineer), an Assembly Statement of Work, a Manufacturing Statement of Work, or other types available from the appropriate Ford organization. See the Global Product Development System (GPDS) for specific timing.

- *APQP*: The External Supplier APQP/PPAP Readiness Assessment (Schedule A) is available through https://web.qpr.ford.com/sta/APQP.html.
  - The organization shall submit completed schedule as specified in the Schedule A notification letter for each program (monthly and after any significant change in APQP status). This applies to priority and non-priority suppliers, see Supplier Engagement Process on https://web.qpr.ford.com/sta/GPDSSupplierEngagement.html.
  - Even if the organization has not received a Schedule A notification letter for a program, but has New Tooled End Items (NTEIs) for a Ford program launch, the organization is still required to complete a Schedule A for each program milestone for all NTEIs and retain the final Schedule A in the PPAP file for the life of part (production and service) plus 1 year.

- *Prototypes*: When the organization is also sourced with the production of prototypes, effective use should be made of data from prototype fabrication to plan the production process. The organization records the dimensional data per the Prototype Build Control Plan, reviews the measured characteristics with Ford PD Engineer and obtains approval on the results from the Ford PD Engineer with confirmed acceptance of parts. If prototype parts are not fully compliant to specification, Ford PD Engineering can approve use of the part with a WERS Alert.
  - The organization should use the APQP/PPAP Evidence Workbook to record prototype part data for Ford PD review.
  - The APQP/PPAP Evidence Workbook is available through https://web.qpr.ford.com/sta/APQP.html.
- *Prototype tooling*: Within 30 days of PV (PPAP Phase 2) completion, the organization shall (a) complete the "Prototype Disposal Request" form, which can be obtained through a request to fordtool@ford.com, (b) submit the completed form to D&R supervisor for signature concurrence, and (c) submit signed form to fordtool@ford.com for processing

8.1.1–8.2 No additional requirements.

8.2.1 *Customer communication:* After part approval, the organization shall use the SREA (Supplier Request for Engineering Approval) process to submit approval requests for organization-initiated process change proposals. See https://web.qpr.ford.com/sta/SREA.html.

8.2.1.1–8.2.3.1 See ISO 9001:2015 requirements.

No additional requirements.

8.2.3.1.1 *Review of the requirements for products and services – supplemental*: The customer authorization for waiving formal review may be obtained from the appropriate Ford Organization (Ford Engineering, Purchasing, etc.).

8.2.3.1.2 *Customer-designated special characteristics symbols*: The organization is to contact Ford Engineering to obtain concurrence for the use of Ford Motor Company special characteristics symbols defined as

1. *Critical characteristic*: CC or with safety or legal consideration, ∇
2. *Significant characteristic*: SC – Not relating to safety or legal considerations
3. *High impact (HI) characteristics*: None
4. *Operator safety characteristics (OS):* For internal use, the organization may develop its own special characteristics symbols.

The special characteristics definitions are available in the *Ford FMEA Handbook* (2011).

Ford Designated Special Characteristics.

- *Critical characteristic (∇) parts*: Ford designated Control Item Parts are selected products identified by Ford Engineering, concurred by Ford/organization manufacturing and identified on drawings and specifications with an inverted delta (∇) preceding the part. Control Item products have critical characteristics that may affect safe vehicle operation and/or compliance with government regulations. Unique symbols identify safety and regulatory characteristics on components equivalent to the inverted delta (∇) symbol.

- *Fasteners with critical characteristics*: For fasteners, base part numbers beginning with "W9" are to be treated as inverted delta. Critical characteristics for fasteners may be designated by methods defined in Ford Engineering Fastener Specifications available through Ford Global Materials and Fastener Standards. Other special characteristics – Significant and High Impact and Operator Safety Characteristics – are described in the *Ford FMEA Handbook*.

8.2.3.1.3 *Organization manufacturing feasibility:* Manufacturing feasibility reviews for updated or new manufacturing processes or capacity increases requiring tooling or equipment shall be documented as specified on the Manufacturing Feasibility form (both initial feasibility and final feasibility) (https://web.qpr. ford.com/sta/Feasibility_Form.xls) per the timing specified on https://web.qpr. ford.com/sta/APQP.html and shall include all appropriate organization and Ford organizations.

8.2.3.2–8.3.1.1 See ISO 9001:2015 requirements.

No additional requirements.

8.3.1.1 *Design and development of products and services – supplemental*: The organization should consider Incoming Inspection when developing control strategies to prevent the use of non-conforming incoming material.

8.3.2 See ISO 9001:2015 requirements.

8.3.2.1 *Design and development planning – supplemental. FMEA and Control Plan Development*: FMEAs and Controls Plans shall ensure that the manufacturing process complies with Critical to Quality process requirements as specified in the Supplier Manufacturing Health Charts located at https://web.qpr.ford.com/sta/Supplier_Manufacturing_Health_Charts.html.

Approvals required for Inverted Delta parts:

- Process FMEA(s) and Control plan(s) for inverted delta component(s) require Ford Engineering & STA approval in writing.
- Approvals required for all parts where the organization is design responsible
- Design FMEA(s) prepared by design responsible organizations requires Ford Engineering approval in writing.
- Approval of revisions to these documents after initial acceptance per the above is also required.
- Ford reserves the right to require approval of FMEA and/or control plans for any part from any organization.

FMEA requirements: Organizations shall comply with the *Ford FMEA Handbook* (2011) requirements – see FSP Document Library https://fsp.portal.covisint.com/web/portal/document_library. Organizations complying with the *Ford FMEA*

Handbook will meet the FMEA and related requirements of the Q1 Manufacturing Site Assessment. [FMEAs following the AIAG/VDA protocol are also acceptable.]

- *Families of FMEAs*: The organization may write FMEAs for families of parts, where typically the only difference in the parts is dimensional, not form, application or function. The organization should obtain STA review and concurrence prior to use of family process FMEAs. The organization should obtain Ford PD review and concurrence prior to use of family design FMEAs.
- *FMEA documentation*: Organizations are to provide copies of FMEA documents to Ford Motor Company upon request.
- *Special Characteristic traceability for build to print organizations*: For build to print organizations, the organization shall obtain from Ford DFMEA information (including potential Critical Characteristics – YCs and potential Significant Characteristics – YSs) to develop the PFMEA and special characteristics (CC, SC, HI, and OS, as appropriate). The organization shall document special characteristics on the Special Characteristics Communication and Agreement Form – SCCAF (FAF03–111–2) including where special characteristics are controlled at sub-tier suppliers and obtain Ford approval. The SCCAF template is available through APQP/PPAP Evidence Workbook (through https://web.qpr.ford.com/sta/APQP.html). This also applies to Ford-directed sub-tier suppliers without a Multi-Party Agreement.
- *Documentation of Controls for Critical Characteristics*: Both build-to-print and design responsible organizations identify in the APQP/PPAP Evidence Workbook the special controls to prevent shipment of any nonconformance to Ford specified Critical Characteristics, regardless of the location of the special controls in the supply chain (tier 1 through tier N).
- *Control Plans*: All Ford Motor Company parts shall have Control Plans (or Dynamic Control Plans – DCP if required by Powertrain).
- *Special Characteristic Traceability*: Special Characteristics and control approach are traceable from the DFMEA through the PFMEA, process flow chart and the SCCAF to the Control Plan and recorded in the APQP/PPAP Evidence Workbook.
- *Ongoing Engineering Specification testing documentation*: Any revisions to the Product Validation Engineering Specification or other inspection frequencies in the Control Plans and PFMEAs require Ford approval through the Supplier Request for Engineering Approval (SREA).
- *Pre-Launch Control Plans*: Pre-Launch Control Plans shall be completed and utilized during production of parts from <TT>/<Unit TT> until final process capability approval is achieved. **Note**: The Production Control plan may be used for demonstration of Phase 3 with STA concurrence
- *Submission of Pre-launch Control Plan Data*: Organizations providing parts to Ford Powertrain plants shall submit, to the Ford Powertrain Plant,

the Pre-launch Control Plan data for all <Unit TT> and <Unit PP> parts as specified by Ford.

- *Control Item* (∇) *Fasteners*: The following control shall be included in the Control Plan for fasteners that are Control Items:
  - *Material Analysis – Heat-Treated Parts*: Prior to release of metal from an identified mill heat, a sample from at least one coil or bundle of wire, rod, strip, or sheet steel shall be analyzed and tested to determine its conformance to specifications for chemical composition and quenched hardness. The organization shall test a sample from each additional coil or bundle in the heat for either chemical composition or quenched hardness. The organization shall document the results and include the steel supplier's mill heat number. This requirement applies to both purchased material and material produced by the organization.
  - *Material Analysis – Non-heat-Treated Parts*: The organization shall visually check the identification of each coil or bundle of wire, rod, strip, or sheet steel to determine that the mill heat number agrees with the steel supplier's mill analysis document and applicable specifications. The organization shall test each coil or bundle for hardness and other applicable physical properties.
- *Lot Traceability*: The organization shall maintain lot traceability.

8.3.2.2–8.3.3.3 See ISO 9001:2015 requirements. No additional requirements.

See also 8.2.3.1.2 for Ford customer-specific requirement regarding customer-defined symbols.

8.3.4 *Design and development controls*: The organization shall perform DV to show conformance to the appropriate Ford Engineering requirements: Attribute Requirements List (ARL) and System Design Specification (SDS). The organization shall record the DV methods with the test results and submit to Ford Product Engineering for approval.

- For organizations responsible for component level DV testing, the organization shall have a documented Design Verification Plan and Report (DVP&R) that includes organization/sub-tier supplier and Ford responsible test(s) as applicable. The organization provides evidence of successful completion on all component level DV testing on the DVP&R. The organization shall obtain Ford PD engineer approval for all tests and results. These requirements apply to all organizations; regardless of the organization's or part's PPAP submission level or design responsibility.
- ARLs and SDSs are available from Ford Product Engineering: The organization shall use GPDS (Global Product Development System) when reviewing product design and development stages. Information on GPDS is available through FSP (Ford Supplier Portal https://fsp.covisint.com); log in to Ford Supplier Portal and then go to the Ford Supplier Learning Institute (FSLI) application.

- *Product development*: For Inverted Delta (∇) parts, design responsible organizations shall include Ford Engineering and Assembly/Manufacturing in GPDS milestone design reviews, as appropriate.
- Where feasible, design responsible organizations shall include Ford Engineering and Ford Assembly and/or Manufacturing in design reviews for all Ford parts.

8.3.4.1–8.3.4.2 No additional requirements

8.3.4.3 *Prototype program*: The organization is responsible for the quality of the parts it produces and for any subcontracted services, including sub-tier suppliers specified by Ford Motor Company without a Multi-Party Agreement. This applies to all phases of product development, including prototypes. Individual Statements of Work may specify alternate responsibilities. See GPDS for additional information on prototype programs on Ford Supplier Portal.

8.3.4.4 *Product approval process production part approval process*: For production parts and approval of components from sub-tier suppliers, the organization shall comply with the AIAG Production Part Approval Process (PPAP) manual and Ford's Global Phased PPAP available through https://web.qpr.ford.com/sta/Phased_PPAP.html. Additional requirements are specified in Q1 https://web.qpr.ford.com/sta/Q1.html.

- For service parts, in addition to meeting the requirements of the AIAG Production Part Approval Process (PPAP) manual, the organization must comply with the AIAG Service Production Part Approval Process (Service PPAP) manual.
- Submission of sub-tier supplier PPAP: Evidence of sub-tier component part approvals may be a summary (approved PSWs, a listing of PSW approvals or equivalent).
- Organization initiated changes: Per PPAP, the organization shall submit via WERS all organization-initiated design change proposals, unless the organization or sub-tier supplier does not have access to WERS.
- After SREA approval and change implementation, all changes require PPAP approval and functional trial approval or PPAP approval and functional trial waiver prior to shipping production quantities.
- STA will not grant full PPAP approval if the part or manufacturing process is under WERS Alert, per exception management process. See https://web.qpr.ford.com/sta/Phased_PPAP.html.
- SREAs for service parts: The organization should process supplier-initiated change requests associated with Service-Unique parts no longer used in Ford production via the applicable FCSD Service Part Deviation SREA process found via https://web.srea.ford.com/ through the Ford Supplier Portal. Contact your local FCSD STA engineer for further clarification.

8.3.5 See ISO 9001:2015 requirements. No additional requirements

8.3.5.1 *Design and development outputs – supplemental*: Assistance in C3P or legacy data system compatibility with Ford CAD systems is available through https://web.c3p.ford.com/index.html

8.3.5.2–8.4.1.1 See ISO 9001:2015 requirements.

No additional requirements.

8.4.1.2 *Supplier selection process*: The organization's supplier selection process should include evaluation of the supplier's supply chain management system. The organization shall complete a financial assessment of the supply chain at a minimum annually, in conjunction with the annual audit program (see 9.2.2.2 of IATF 16949), not just at the initial supplier selection.

8.4.1.3 *Customer-directed sources (also known as "Directed–Buy")*: When required by the contract with Ford, the organization shall obtain approval from Ford Motor Company prior to sourcing sub-tier suppliers. Please contact the Ford Buyer.

8.4.2 See ISO 9001:2015 requirements.

No additional requirements.

8.4.2.1 *Type and extent of control – supplemental*: The organization shall have incoming product quality measures and shall use those measures as key indicators of sub-tier supplier product quality management.

8.4.2.2 *Statutory and regulatory requirements*: Applicable regulations shall include international requirements for export vehicles as specified by Ford Motor Company, e.g., plastic part marking (E-4 drafting standard – WSS-M99P9999-A1 and European End of Life of Vehicle (ELV) – available on FSP (Ford Supplier Portal https://fsp.covisint.com). Material reporting requirements for ELV are specified by WSS-M99P9999-A1 under "Important Documents.")

8.4.2.3 *Supplier quality management system development*: The organization may meet this requirement by successful assessments of the sub-tier suppliers per the authorization stated on https://web.qpr.ford.com/sta/. The frequency of these reviews shall be appropriate to the sub-tier supplier impact on customer satisfaction.

Sub-tier supplier quality management system requirements:

- Where a sub-tier supplier is not third-party certified to IATF 16949, Ford reserves the right to require the organization to ensure sub-tier supplier compliance with the "Minimum Automotive Quality Management System Requirements for Sub-tier Suppliers" available through http://iatfglobaloversight.org/default.aspx. Evidence of effectiveness shall be based on having a defined process and implementation of the process including measurement and monitoring.
- Where any organization has sub-tier suppliers not third-party certified to IATF 16949, the organization is encouraged to require sub-tier supplier compliance with the "Minimum Automotive Quality Management System Requirements for Sub-tier Suppliers."
- Ford or organization second-party assessment or third-party certification of sub-tier suppliers does not relieve the organization of full responsibility for the quality of supplied product from the sub-tier supplier (including Ford-directed sub-tier suppliers without a Multi-Party Agreement).

- Although all sub-tier suppliers must be assessed per this section, sub-tier supplier improvement efforts shall focus on those sub-tier suppliers with the highest impact on Supplier Improvement Metrics (SIM).
- *Sub-tier supplier management process*: Organizations are encouraged to apply the principles outlined in "CQI-19 AIAG Sub-tier Supplier Management Process Guideline" to all their sub-tier suppliers.
- Additionally, Ford reserves the right to require the organization to apply the principles outlined in "CQI-19 AIAG Sub-tier Supplier Management Process Guideline" to address issues identified in the organization's supplier development and management process. Ford will communicate the requirement to apply CQI19 to the specifically selected organization(s) based on sub-tier supplier management issues attributed to the organization. Evidence of effectiveness shall be based on having a defined process and implementation of the process including measurement and monitoring.
- *Critical Characteristic Controls at the Sub-tier Suppliers*: For Critical Characteristics, the responsible organization ensures that sub-tier suppliers have controls in place to prevent shipment of nonconforming product at the location where the associated physical characteristics are manufactured by sub-tier suppliers. The sub-tier supplier controls for the Critical Characteristics are identified by the organization in the APQP/PPAP Evidence Workbook. This also applies to Ford-directed sub-tier suppliers without a Multi-Party Agreement.

8.4.2.3.1 No additional requirements.

8.4.2.4 *Supplier monitoring*: In support of Ford's expectation of 100% on-time delivery, the organization shall also require 100% on-time delivery from sub-tier suppliers. The organization shall communicate any delay or risk to the affected Ford customer. The organization should monitor and minimize any premium freight expenses related to sub-tier suppliers for late deliveries. These also apply to Ford-directed sub-tier suppliers without a Multi-Party Agreement.

8.4.2.4.1–8.5.1.1 See ISO 9001:2015 requirements.

No additional requirements

8.5.1.2 *Standardized work – operator instructions and visual standards operators shall use the most current work instructions*: The organization shall ensure that work instructions contain reaction plans for nonconformances showing the specific required steps.

8.5.1.3–8.5.1.7 See ISO 9001:2015 requirements.

No additional requirements

8.5.2 *Identification and traceability*: The organization shall meet all logistics requirements as specified by Material Planning and Logistics (MP&L). MP&L requirements are available in the Global Terms & Condition (GTC) web guides at https://web.fsp.ford.com/gtc/production/index.jsp?category=guides and on MP&L-in-a-Box at https://comm.extsp.ford.com/sites/MPLB2B/Pages/MPLdefault.aspx. The organization is required to achieve level "A" on the Material Management Operation Guideline/Logistics Evaluation (MMOG/LE) to achieve and maintain

Q1. Key requirements for MMOG/LE (Material Management Operation Guideline/Logistics Evaluation) compliance include

- Annual MMOG/LE assessment completed and reported May 1 to July 31 each year
- Adherence to Ford production and service delivery rating requirements for all regions as stated in Q1
- Part identification and tracking
- Lot traceability throughout the value chain (lot traceability shall include subcontracted components of an assembly/module that are associated with compliance to any inverted delta requirement)
- Electronic communication with Ford and sub-tier suppliers
- Management and maintenance of the Ford DDL CMMS3 system
- Prevention of damage or deterioration of supplied products
- Use of the appropriate packaging forms and maintenance of the Ford DDL CMMS3 DAIA Packaging screen, as applicable. Packaging requirements and forms can be found in the packaging GTC Web Guides at https://web.fsp.ford.com/gtc/production/index.jsp?category=guides
- Management and maintenance of returnable dunnage. Returnable container requirements are available through the GTC Web Guides at https://web.fsp.ford.com/gtc/production/index.jsp?category=guides
- Adequately trained personnel, as defined in MMOG/LE
- In all cases, if unsure of the MP&L requirements, contact the production and service delivery analyst for the organization site, for each region. The analyst contact information is available through SIM.

Inverted delta part identification: The inverted delta symbol (∇) shall precede the Ford Motor Company part number for parts with Critical Characteristics, in accordance with the Packaging Guidelines for Production Parts and Shipping Parts/Identification Label Standard, both available through Ford Supplier Portal MP&L page https://comm.extsp.ford.com/sites/MPLB2B/Pages/MPLdefault.aspx

Note: Branding (E108) does not require the inverted delta symbol to be included with the part number physically marked on the part.

8.5.2.1–8.6 See ISO 9001:2015 requirements.

No additional requirements.

8.6.1 *Release of products and services – supplemental*: Ford reserves the right to require the use of an independent third-party inspector to ensure that the organization only ships compliant product to Ford facilities.

8.6.2 *Layout inspection and functional testing*: The organization shall perform annually a layout inspection (to all dimensional requirements) on at least five parts. Where tooling has multiple cavities, tools, or centers, the organization conducts the annual layout on at least one part from each cavity, tool, or center, with a minimum overall sample of five parts.

Note: Five parts are not required from each cavity; tool or center, only a minimum of one part is required from each cavity, tool, or center. The measurements are to be documented on the APQP/PPAP Evidence Workbook (see the "Prototype or

Production Measurement Results" section), available through. https://web.qpr.ford.com/sta/APQP.html.

8.6.3 *Appearance items*: Appearance approval requirements are specified in PPAP, Ford customer specific requirements https://web.qpr.ford.com/sta/Phased_PPAP.html.

8.6.4–8.6.5 See ISO 9001:2015 requirements.

No additional requirements.

8.6.6 *Acceptance criteria*: For guidance on product monitoring and reaction plan techniques for product conformance to specification, see the references AIAG SPC and APQP.

For ongoing process capability requirements, see Table 3.2 in 9.1.1.2 clause.

8.7–8.7.1 See ISO 9001:2015 requirements.

No additional requirements.

8.7.1.1 *Customer authorization for concession*: Ford Motor Company authorization of product differing from Ford specifications is managed by Worldwide Engineering Release System (WERS), limited to the quantity of parts or time-period approved in the WERS Alert. This is applicable to both prototype- and production-level parts.

- PPAP submission and Interim PSW acceptance are required for production use of parts with a WERS Alert. Where written by the organization, Alerts must contain the following:
  - The specific PPAP requirements that are not completed
  - The modified specifications(s) that the part satisfies
  - The justification why the modified specification(s) is acceptable
  - The containment plan to assure the quality of parts (e.g., extraordinary controls/inspection process/robust measurement systems).
- The period (typically in terms of days), the number of parts, and the specific launch build event for which the Alert is effective.
- The WERS help desk can provide information on WERS via email: hwers@ford.com.
- WERS training is available through Ford Supplier Learning Institute (FSLI) through https://fsp.portal.covisint.com/web/portal/home.

8.7.1.2–8.7.2 See ISO 9001:2015 requirements.

No additional requirements.

Clause 9: Performance Evaluation

9.1 *Monitoring, measurement, analysis, and evaluation*: Ford reserves the right to request the data collected by the organization as defined in either the pre-launch or production control plans.

9.1.1 See ISO 9001:2015 requirements.

No additional requirements

9.1.1.1 *Monitoring and measurement of manufacturing processes:* In Table 3.2 of this clause details, the ongoing process capability requirements. All process controls shall have a goal of reduction of variability, using six-sigma or other

TABLE 3.2
Ongoing Process and Product Monitoring

The Control Chart Indicates That the Process	Actions on the Process Output Based on Process Capability (Ppk)	
	<1.33	≥1.33
Is in control	100% inspect[a]	Accept product
		Continue to reduce product variation
Has gone out of control	*Identify special cause*	
	100% inspect[a] all product since the last in-control sample	

[a] The organization ensures that the 100% inspection methodology prevents shipment of any nonconforming product to Ford. The 100% inspection methodology would typically include error-proofing, such as a poka-yoke.

The organization ensures that Critical Characteristics (CC) have controls which prevent the shipment of nonconforming product, regardless of the location in the supply chain (tier 1 through tier N) of the manufacture of the physical characteristic(s) associated with the Critical Characteristic. The organization records the CC controls in the APQP/PPAP Evidence Workbook.

Statistical process control on product characteristics without continuous manufacturing process controls is not appropriate or sufficient for Critical Characteristics.

appropriate methods. The Statistical Process Control Manual in Section 2.11 provides additional guidance where tool wear impacts variability. All process metrics are to be traceable to Ford requirements.

9.1.1.2 *Identification of statistical tools*: The organization shall use the latest edition of the following references as appropriate: See IATF 16949 for applicable references

Process capability: The capability index for reporting launch process capability and ongoing production process capability is Ppk (Performance Index). See Ford's PPAP customer specifics for the launch process capability requirements. https://web.qpr.ford.com/sta/Ford_Specifics_for_PPAP.pdf. See Table 3.2 for ongoing process capability requirements. The organization shall maintain ongoing process capability at Ppk ≥ 1.33. The requirement for maintenance of ongoing process capability is to be included in the production Control Plan and the capability results recorded in the APQP/PPAP Evidence Workbook. The results of monitoring process capability are to be available to Ford upon request. When investigating a process capability issue, it is advisable to use multiple indices, e.g., Pp, Ppk, Cp, and Cpk. When used together, the indices assist in the determination of sources of variation (see the references on Statistical Process Control).

9.1.1.3 No additional requirements

9.1.2 *Customer satisfaction*: The organization shall monitor performance and customer satisfaction metrics (as defined by Q1) and updates to Ford requirements on Ford Supplier Portal (FSP) https://fsp.covisint.com. It is recommended that the organization review their performance status on Supplier Improvement Metrics

(SIM) at least weekly. (Some information is updated daily in SIM.) At least twice per year, the organization shall communicate customer satisfaction metrics to all employees who affect the quality of Ford Motor Company parts.

9.1.2.1 *Customer satisfaction – supplemental Certification Body Notification*: The organization shall notify its CB of record in writing within 5 working days if Ford Motor Company places the site on Q1 Revocation. This notification of the CB will constitute a "customer claim" as defined by the IATF 16949 Rules. This step will suspend the organization's IATF 16949 certification. However, a suspended certification is still acceptable for Q1 Capable Systems requirements. Even though the CB may request a status report from Ford on the implementation of corrective actions to address the Q1 revocation, the CB alone must determine whether to withdraw the IATF 16949 certification within 90 days of suspension.

Note 1: Reinstatement of Q1 from Revocation requires at least 6 months of acceptable organization performance. The CB may remove suspension of the IATF 16949 certificate if the organization's corrective actions have addressed the nonconformances leading to the certificate suspension, as determined by the CB. The CB may remove the suspension even though the site remains under Q1 Revoked status, accumulating the required 6 months of acceptable performance data.

Note 2: At its option, Ford may file an OEM performance complaint with a CB when confronted with a specific organization quality performance issue where a root cause may be a nonconformance in the organization's quality management system. Ford will send the notification letter to both the organization and the Certification Body's Oversight Office.

9.1.3–9.2.2.2 See ISO 9001:2015 requirements.

No additional requirements

9.2.2.3 *Manufacturing process audit Ford manufacturing*: Process Assessment Requirements. The organization is responsible to ensure that all tiers of suppliers are assessed to the applicable Ford manufacturing process standards.

Note: Self-assessment by the sub-tier suppliers, including implementation of corrective action plans as required, meets this requirement.

Refer to https://web.qpr.ford.com/sta/Ford_GTS.html on Ford Supplier Portal for all these standards except AIAG CQI-xx, which are available through AIAG.

- *Ford Supplier Manufacturing Health Chart Requirements*: The organization shall assess compliance to Critical to Quality process requirements in accordance with APQP as specified in the Supplier Manufacturing Health Charts located at https://web.qpr.ford.com/sta/Supplier_Manufacturing_Health_Charts.html.
- *Heat Treat Assessment Requirements*: Organizations and sub-tier suppliers providing heat treated product and heat-treating services shall demonstrate compliance to AIAG CQI-9 "Special Process: Heat Treat System Assessment" and Ford Specific CQI-9 requirements (available through https://web.qpr.ford.com/sta/CQI9_Ford_Specific_requirements.xls); CQI-9 is available through AIAG http://www.aiag.org/.
- *CQI-9 Special Process*: Heat Treat System Assessment. All heat-treating processes at each organization and sub-tier supplier manufacturing site

shall be assessed annually (at all tier levels), using the AIAG CQI-9 "Special Process: Heat Treat System Assessment" (HTSA) and Ford. Specific CQI-9 requirements. Assessments are also required following any heat treat process and/or changes of heat treat equipment or additions. The organization must review that the individual assessments are current (<12 months old), meet the requirements above, and enter the CQI-9 assessment status into GSDB Online.

- GSDB Online is accessible through: https://web.gsdb2.ford.com/GSDBeans/servlet/gsdbeans.web.lib.GSDB.
- The organization shall maintain the 2-prior annual CQI-9 assessment reports and related information at the organization's site and make them available to STA upon request. Heat Treat assessments are conducted by the organization, heat treat suppliers, sub-tier suppliers or by Ford. Demonstration of compliance to CQI-9 and Ford Specific CQI-9 requirements does not relieve the organization of full responsibility for the quality of supplied product.
- To reduce the risk of embrittlement, heat-treated steel components shall conform to the requirements of Ford Engineering Material Specification WSS-M99A3-A, also available per Section 0 of this document.

9.2.2.4–9.3.1 See ISO 9001:2015 requirements.

No additional requirements

9.3.1.1 *Management review – supplemental*: The organization management shall hold monthly QOS performance meetings as specified in the Q1 Manufacturing Site Assessment available on https://web.qpr.ford.com/sta/Q1.html. The results of these QOS performance reviews shall be integral to the senior management reviews.

Note: The organization need not hold management review as one meeting, but it may be a series of meetings, covering each of the metrics monthly.

9.3.2 *Management review inputs*: Management review input must also include the Q1 Manufacturing Site Assessment results

9.3.2.1–9.3.3.1 See ISO 9001:2015 requirements. No Ford customer-specific requirement for this section.

Clause 10: Improvement

10.1 See ISO 9001:2015 requirements.

No additional requirements

10.2–10.2.2 The organization shall have processes and systems in place to prevent shipment of non-conforming product to any Ford Motor Company facility. The organization should analyze any nonconforming product or process output using the [G]8D methodology to ensure root cause correction and problem prevention.

- Customer concerns: Organizations shall respond to Quality Rejects (QRs) by

    Note: The clock starts once Ford has sent the notification to the organization
    - Responding in 24 hours

- Implementing containment in the Ford plant. The organization and/or third-party must follow local procedures and site rules while carrying out containment
- Providing certified stock
- Delivering an [G]8D, beginning with Symptom and Emergency Response Actions (D0) through Interim Containment Actions (D3)
- Within 48 hours of notification by the Ford plant, the organization shall notify Ford Service if the product quality issue is suspected of affecting any FCSD shipments
- Within 15 calendar days delivering the [G]8D or (six sigma) six panel with preliminary or verified root cause, and a plan to implement corrective and preventive actions with supporting data. A summary of the Quality Reject Process for North America is available through https://web.qpr.ford.com/sta/QR2NA.htm. Global 8D system is available on FSP (https://web.quality.ford.com/g8d/).
- *Returned product test/analysis*: The organization shall have a documented system for internal notification, analysis, and communication of all Ford plant returns and warranty returned parts.
  - The organization shall communicate the results of analysis to the responsible Ford and organization work groups and include the results in the associated [G]8D report.
  - The organization shall communicate Ford plant PPM (Parts Per Million) to all organization plant team members.
  - The organization shall develop a system to monitor Ford plant and warranty concerns. The organization shall also implement corrective actions to prevent future Ford plant and warranty concerns.
  - Returned product test results are to be included in the monthly QOS performance report as part of the Management Review.

10.2.3–10.2.4 No additional requirements.

10.2.5 *Warranty management systems*: Organizations shall comply with the requirements in the Ford Production Purchasing Global Terms & Conditions (PPGT&Cs), Applicable Web Guides, Q1. Requirements and the Manufacturing Site Assessment https://web.qpr.ford.com/sta/Q1.html.

10.2.6–10.3.1 See ISO 9001:2015 requirements. No additional requirements.

4 Documentation

OVERVIEW

This chapter provides the rationale and some guidelines for the content and format of the quality manual (if it exists), procedures, and instructions. In addition, it explains the mechanics and requirements of design control, control documentation, and the communication of the documentation in the organization.

Procedures are specific to a particular function and are identified and/or referenced in the quality manual if it exists. Instructions, on the other hand, may be parts of the procedures or separate documents. In either case, they represent detailed instructions for the performance of a particular task within a procedure. For more information on instructions, see Wilson (1996), Brumm (1995), and Clements et al. (1995).

ALL ISO standards enable firms involved in international trade to obtain a degree of confidence in the quality of the work done by current and potential suppliers. Furthermore, the assumption on which they operate is that if the process is effective, the product and/or service will more likely than not be of high quality.

This effectiveness, however, involves conformance/compliance with a set of standards and procedures, which the organization has defined based on its products or services. Conformance or compliance, on the other hand, is assured through a formal system of inspection and audits. This formality is addressed in the ISO, industrial, and customer-specific requirements as requirements that a firm must document its processes and conform to the statements of its own process.

Where the earlier certifiable – contractual – standards did not require much in writing, the new editions of the standards require both documentation and maintenance of documentation (procedure and other quality records). It is profoundly important to mention that the new version of the ISO 9001 DOES NOT require a quality manual.

This requirement is very specific in ISO 9001. It is also very important to remember that unless something is in writing or corroborated, it is not considered as "proof" of being documented. In fact, from our experience of performing audits, the problem of not having appropriate documentation is so great that the most common reason for failure during the third-party audit (certification/registration) is related to inadequate documentation. So, let us not forget that the ISO 9001:2015: clause 7.5.1 quality management system documentation states unequivocally that documented information is required. The format and structure of the documentation is at the discretion of the organization and will depend on the organization's size, culture, and complexity. If a series of documents is used, then a list shall be retained of the documents that comprise the quality manual for the organization."

In addition, ISO 9004 Section 5 – *Strategy and policy* brings in assessment of the needs and expectations of other interested parties, not just customer requirements as in ISO 9001. The goal is to look at the organization's whole environment, where ISO 9001 is focused on customers and their requirements.

Therefore, to satisfy these ISO guidelines, there are three basic requirements.

1. *Documentation of specific quality standards and procedures*: This requirement is perhaps the easiest to comply with, since the company defines how they do and what they do. The intent here is to ensure that a firm's way of doing business can be evaluated and follows the ISO standard. The organization is expected to produce a true reflection of the way the firm does business and the way employees and suppliers perform their functions. This fundamental principle of standard conformance/compliance can be summarized as *Say what you do, do what you say you do*. It can also be formalized in the quality manual (if it exists), quality procedures, and instructions.

2. *Documentation of product development life cycle results*: This requirement consists of specifications, design documentation, test plans, and other descriptions of the product or the development and acceptance process. Its purpose is to establish continuity and control during product development, production, and maintenance in the life of the product.

3. *Documentation of the outcomes that are required under the standards*: This requirement represents the verification of outcomes from specific tasks, reviews, audits, inspections, etc. They are the quality records and produce an audit trail that is used to verify that the organization performs the functions described in its standards and procedures documentation. Quality records, in *Do what you say*, generally provide the basis for performance evaluation towards continuous process improvement.

So far, we have addressed the issue of documentation from the standard's perspective. Let us now look at some of the specific benefits that the organization may gain. First, all documentation – regardless of source – should benefit the organization in some way, either directly (internally) or indirectly (externally). Second, all documentation should be as specific as possible in order to guide the organization's continual improvement. For specific benefits, we examine the following components of documentation.

- *Standards and procedures:* If they indeed represent reality (current practice) and are accurate, they will contribute to the firm so as to
  - Make the training of new staff easier.
  - Permit more objective evaluation of performance.
  - Provide continuity and consistency when there is staff turnover. Facilitate training current employees for specific tasks.
  - Provide a benchmark for future improvement.
- *Quality records*: They document an activity and provide the evidence and documentation of conformance/compliance with preestablished standards and/or procedures. In addition, as records accumulate, they provide a database for performance analysis and performance improvement.
- *Life cycle documentation*: It provides continuity across the product development life cycle. In other words: The product life cycle it is a guide that outlines the product development and technical support resources available during a product's life span.

- *Document control*: Document control is a requirement. The idea of document control is that documentation must be accessible to those who have need and authorization to use it and that it must be kept up to date.

So, what are the required documents for the ISO 9001: 2015?

When applying for your ISO 9001 certification, you will submit two groups of documents to the external auditor:

1. The documentation named by the standard (as provided below)
2. The documentation you decide is required for your QMS.

THE MANDATORY DOCUMENTS REQUIRED BY ISO 9001:2015

- Documented information to the extent necessary to have confidence that the processes are being carried out as planned (clause 4.4).
- Evidence of fitness for the purpose of monitoring and measuring resources (clause 7.1.5.1).
- Evidence of the basis used for calibration of the monitoring and measurement resources (when no international or national standards exist) (clause 7.1.5.2).
- Evidence of competence of person(s) doing work under the control of the organization that affects the performance and effectiveness of the QMS (clause 7.2).
- Results of the review and new requirements for the products and services (clause 8.2.3).
- Records needed to demonstrate that design and development requirements have been met (clause 8.3.2).
- Records on design and development inputs (clause 8.3.3).
- Records of the activities of design and development controls (clause 8.3.4).
- Records of design and development outputs (clause 8.3.5).
- Design and development changes, including the results of the review and the authorization of the changes and necessary actions (clause 8.3.6).
- Records of the evaluation, selection, monitoring of performance and re-evaluation of external providers, and any actions arising from these activities (clause 8.4.1).
- Evidence of the unique identification of the outputs when traceability is a requirement (clause 8.5.2).
- Records of the property of the customer or external provider that is lost, damaged, or otherwise found to be unsuitable for use and of its communication to the owner (clause 8.5.3).
- Results of the review of changes for production or service provision, the persons authorizing the change, and necessary actions taken (clause 8.5.6).
- Records of the authorized release of products and services for delivery to the customer including acceptance criteria and traceability to the authorizing person(s) (clause 8.6).

- Records of nonconformities, the actions taken, concessions obtained, and the identification of the authority deciding the action in respect of the non-conformity (clause 8.7).
- Results of the evaluation of the performance and the effectiveness of the QMS (clause 9.1.1).
- Evidence of the implementation of the audit program and the audit results (clause 9.2.2).
- Evidence of the results of management reviews (clause 9.3.3).
- Evidence of the nature of the nonconformities and any subsequent actions taken (clause 10.2.2).
- Results of any corrective action (clause 10.2.2).

These documents need to be retained as records of the results of your QMS. The list described here may look an overburdened one but upon further examination one may find it is not overwhelming, for many companies find these records already exist within the company's current documentation practice. There's a good chance that you will already have or be familiar with most of these documents. Newer businesses (or those new to the ISO 9001:2015 standard) who don't have a long or broad documentation history are the ones who usually spend the most time generating the paperwork listed above.

Ultimately, the documented information is part of the core value of the ISO 9001:2015. It encourages you to standardize the processes you already employ and to work towards consistent data collection and data updates to core paperwork like the documents listed above.

A Guide to the Non-mandatory Documentation

Beyond of the documents listed above, your organization may choose to add documents that contribute to the value of your QMS. According to the ISO, non-mandatory ISO 9001:2015 documentation includes

- Organization charts
- Procedures
- Specifications
- Work instructions
- Test instructions
- Production schedules
- Approved supplier lists
- Quality plans
- Strategic plans
- Quality manuals
- Internal communication documents.

If you do create these types of documented information, then you must follow the same rules laid out in clause 7.5. In other words, treat them the same way as you treat the named required documented information.

WHAT ABOUT THE QUALITY MANUAL?

Those familiar with the ISO 9001:2008 will remember the need to complete and present the quality manual to the certification body before the audit. The publication of the 2015 standard revealed that **the quality manual is no longer a mandatory document**.

If you are certified under the 2008 version, you might let out a cheer. But the truth is that even though it's no longer required, it is still a very helpful document to have. What's more, you still need some type of document that describes your QMS (including scope and process interactions) to send to the auditor, even if you don't call it a quality manual). Your organization's quality manual is a comprehensive look at your organization, your QMS, and the approach to quality management you selected on the development journey.

In the past, organizations spent a huge amount of time and energy creating these documents. While most met the fundamental requirements of the ISO 9001, many missed the spirit of the document itself. Instead, they focused too heavily on over-writing the document rather than creating something useful. If you created a quality manual in the past, there's a good chance that you never used it again.

As we already said the Quality Manual (QM) is no longer a mandatory document, according to the new version of the ISO 9001:2015 standard. How did that happen? The Quality Manual was the staple document that a certification body asked for before the certification audit. How has it suddenly lost its purpose and importance?

Well, practically it lost its significance because "most, if not all" used it as a PR document as opposed to focus on the real essence of the organization. No one really read it – especially the people who actually did the work. However, it was a first step for the auditor of the certification body to review the QMS of the organization. By reviewing the QM, the auditor gained a much better understanding of the organization's QMS and became more effective in the auditing process. On the other hand, the QM was a document that demonstrated management's knowledge of what the QM system for their organization was and how it could be managed effectively.

It will be a disaster if any organization's management decides that the QM is not necessary, just because the standard does not call for it. The fact of the matter, however, is that with or without a Quality Manual, organizations will still need some overall QMS document. There will still be a need to send the certification body a document that will describe your system, as well as sending it to big clients. Although it's no longer mandatory, all requirements from the Quality Manual (except 4.2.2b) remain in the new version of the standard. The scope of the QMS and interactions between the processes still need to be defined. These requirements are even more detailed in the new version, and they still must be in some form of documented information.

The new version of the standard has some new requirements that need to be met as documented information, which can be easily included in a Quality Manual – for example, the context of the organization.

So, the question is: Is it really true that there is no need for a formal QM? It depends on how one defines "no need." The truth is *whereas there is no need for a formal QM, the information on a typical QM must be available in some other way.*

Therefore, one may say that the same ideas presented in any QM, now they will be presented in a different form and as a consequence the information will be available.

Again, this new non-mandatory document (or whatever we call it) will replace the Quality Manual and will contain all remaining requirements from clause 4.2.2, and add some new ones.

This new document should provide the following information about the organization:

- We are XYZ company.
- We are producing this and providing these services.
- We apply a quality management system to these processes.
- We don't apply these clauses of the standard for these reasons.
- These are our processes and their interactions.
- This is the internal and external context in which we operate.

This list obviously is an oversimplification, and of course, all this information can't be placed in just one paragraph, but this document would make sense and it would meet most of the requirements from clause 4 of the new version of the standard. Additionally, the organization's mission and vision can be added, and this document can effectively become the brochure that will introduce your company to future clients.

In the final analysis, only time will tell if the new approach will be effective and reasonable for most organizations. To be sure, it is more flexible than the old version and it definitely provides the organization with much more flexibility for innovation and adaptation to its needs.

THE ADVANTAGES OF CREATING QUALITY MANUAL

Even though it's no longer required, producing a quality manual can be incredibly helpful for your organization. It doesn't need to be 50+ pages: it should be short, snappy, and to the point with the goal of contributing to the QMS. In addition, you may find that some businesses require their suppliers to provide their quality manual during the bidding (tender) or selection process. If you have a solid document to provide them, you'll set your organization apart from the competition.

HOW TO SUBMIT AND MAINTAIN YOUR REQUIRED DOCUMENTATION

The ISO 9001:2015 standards aren't prescriptive in terms of the format of your documentation. You can submit paper documents, but you're also able to submit electronic versions of the documents. You can even use video and audio, if you choose. Just remember that when setting up your documentation system, it is helpful to do so in a way that supports the maintenance of and retention of the documents. It should be easy to update, save, (retrieve in any form), and share the documents as and when necessary. Obviously, choosing electronic documentation can help make the process of meeting clause 7.5 requirements, as dictated below.

Clause 7.5: When to review your ISO 9001:2015 documentation is identified in the standard itself. The ISO 9001:2015 requires that you control your documents, but it grants you much more freedom in doing so than the previous 2008 standard did. However, there are still requirements for updating the documented information. Clause 7.5 requires you establish and use documented procedures to "maintain documented information to the extent necessary to support the operation of processes and retain documented information to the extent necessary to have confidence that the processes are being carried out as planned."

For some businesses, the "extent necessary" may mean once a year. Others may only choose to go through the process once every 2 or 3 years. So, when working through your obligation under clause 7.5, you need processes to

- Approve documents
- Review, update, and submit documents for re-approval
- Identify changes
- Make documents available
- Ensure documents are legible and identifiable
- Identify and control external documents
- Keep obsolete documents out of circulation
- Identify obsolete documents as necessary if retained.

You must also establish documented procedures for the following tasks:

- Identifying records
- Storing records
- Protecting records (including keeping them identifiable and legible)
- Retrieving records
- Retaining records
- Disposing of records.

Therefore, one may conclude that the two most important objectives of the ISO 9001:2015 update were to develop a more simple set of standards that apply to all organizations and to allow organizations to focus on the most relevant documentation for their business activities.

Many organizations should already recognize most of the documented information from the list of requirements above (Some of the items on the list have been adopted from https://safesitehq.com/iso-9001-documentation/. Retrieved on April 20, 2020.) The goal then is to standardize it and supplement it with the other documented information that will help your QMS work.

STRUCTURE OF THE DOCUMENTS

The structure of documentation is as important as its content. The structure is the way the information is organized. It enables readers to find information easily and to pinpoint what they need. Structure can be examined from the following points of view:

- Level of detail within subject areas.
- Parsing, based on content, into subject areas.
- Both points of view are legitimate and appropriate. Their use depends upon the organization and the writer.

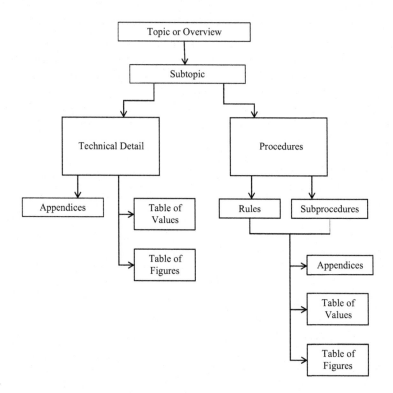

FIGURE 4.1   A typical hierarchical structure.

If the documentation is expected to be procedural or technical in nature, a hierarchical structure is effective and highly recommended. In a hierarchical structure, the information is presented in levels of detail. It starts out with an overview and then branches into one or more levels. A typical hierarchical structure is shown in Figure 4.1.

The overview is the road map of the detailed information and should allow the reader to select the specific part of the topic that is of interest or to skip the topic entirely. An example of an overview in procedures is a *process flow diagram*. Overviews are encouraged, and their frequency or length depends on the complexity or levels of the specific documentation.

The first level of detail is usually the material that contains the main substance of the topic. In a procedure, it is the specific steps to be taken so that the procedure's objective is accomplished. When the levels of detail become cumbersome, they should be put in an appendix or some other reference areas.

With the proliferation of computer technology, documentation structure can be fun and easy to design. With computer-generated documentation, you can chain or cross-reference your documentation, depending on the software and the ability of the user. Some commercial software programs for writing documentation are Quark Xprcss®, HyperText, PageMaker®, WordPerfect®, Excel®, Microsoft®, PowerPoint®, and many others. For a visual representation of computer-generated structure design and a typical content hierarch, see Stamatis (1996, p. 15).

A COMPUTER-GENERATED STRUCTURE DESIGN

While the hierarchical structure defines the level of complexity, parsing (content) is a way of separating things into logical groupings. It is a very subjective activity and care should be taken to make sure that the parsing really reflects common ground of the areas involved. An example of parsing is when information is grouped by volumes, chapters, sections, etc. In procedure writing, a given procedure may be parsed based on the type of work, the performer, the time in the process, or alphabetically by the title of the procedure. For a pictorial view, see Figure 4.2.

The following list is an explanation of its components:

- A volume is a bound document (paper document), and it is generally no more than 1 1/2 inches thick. It is defined by the user, time and frequency of use, and the level of detail. Whenever possible, use only one volume. Volumes may be collected and bound so that they are kept together.
- An appendix is standalone information that supplements but is not essential to the main information in the chapters, sections, or subsections.
- Chapters, sections, and subsections are based strictly on the logic of the material. The deciding factor should be the need and convenience of the reader.
  - A chapter is a combination of sections.

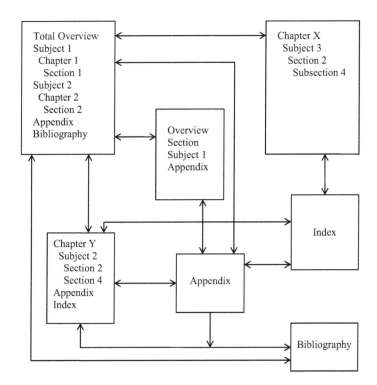

FIGURE 4.2 Computer-generated structure design.

- A section is a labeled set of paragraphs, graphics, or other elements that fully address a topic. A section may also be a combination of subsections.
- A subsection is one to three paragraphs on a specific topic.
- An informational element is the finest level of detail that expresses a complete piece of information relevant to the topic. Usually, it is found as a definition, a step procedure, a description of a part or component, or sometimes as an overview of a topic. An informational element may also be a subsection at the lowest level of the hierarchy.

The discussion of hierarchical structures and parsing may be a bit frightening to a novice. However, there are some basic rules and guidelines that can help you get started.

There is no single correct approach for the documentation structure. Both the hierarchy and the parsing methods are acceptable. The selection process is highly subjective and a matter of preference. For example, for procedural documentation, the flow of the actual procedure is the most common basis of the breakdown.

- Hierarchy and parsing in documentation make writing easier because the writer can divide the document into small chunks and write each one independently. In fact, it is this attribute that allows several people to write parts of the same document. In addition, reading is easier because the information is presented in small chunks. Furthermore, the reader is advised of the content of a section before reading it, by the use of headings and subtitles, which facilitate comprehension.
- When you are ready to write any form of documentation, start with an outline that establishes the contents of the document. The outline (the content hierarchy) will keep you on track. It is always possible to change the outline during the writing process.
- The table of contents permits the reader to find topics and to get an overview of the document's contents. Sometimes, the table of contents is divided into sections and subsections to facilitate finding a particular topic. Generally, the table of contents is written after the completion of the document.
- Indexing is essential in any form of documentation. Indexing is a cross-reference mechanism that locates subjects across the content hierarchy. The structure and the detail of the index is the writer's choice. However, the writer should keep in mind the reader's need of cross-references to find information. Generally, the index is written after the completion of the entire documentation.

QUALITY SYSTEM OVERVIEW

A quality system is the set of policies and procedures that represent the way an organization performs its process(es). The quality system assures that a quality program exists and is followed so that the quality of the product or service is delivered to the customer without problems or nonconformances.

The content of the quality documentation is derived from an analysis of the quality system. Quality events are identified through procedures analysis, typically using a process flow diagram. To generate the content, one must go to the source of the task – use the Gemba approach to observation. The people doing the work know what they do, and they are indeed the best source of information. Therefore, allowing the workers to be active participants in the writing and validation of the quality documentation is a means of ensuring that the documentation accurately reflects the way the firm operates. In addition, it allows employees to have ownership of the documentation, which helps promote quality and increases the intrinsic motivation of the individuals involved.

The objective of all quality documentation is to reflect accurately the way the organization operates. It should describe the current baseline and not the way things ought to be. In addition, the documentation must always be up to date, even though things change. The ISO 9001:2015 edition includes specific clues in the notes section of the standard (non-auditable and non-certifiable items) to help in the interpretation and the implementation of the standards.

THE STRUCTURE OF DOCUMENTATION

A typical documentation of an organization may be structured to meet the requirements of the ISO 9001, ISO 14001, and other international (ISO) standards in a standardized form. The records and forms on the base of the triangle are only representative departments of the organization. They can be arranged to suit to the organization. What is important here is the notion of the involvement of the whole organization.

Even though a generic form accounts for four layers, not all organizations require the same divisions. The actual layers will depend on the complexity of the organization and its products or services. However, the following layers are quite common in most organizations:

1. *Quality manual*: Sometimes called the management manual, operations manual, or policy manual. The quality manual is the highest level of documentation and represents the policy of the organization. Even though it is not a mandatory document for the new ISO 9001, it is a good documentation document to have.
2. *Procedures*: The procedures define the flow of the task or operation. They serve as an overview of the process, job, or task.
3. *Job/work instructions*: Sometimes, these instructions are called standard operating procedures (SOPs). Job/work instructions define how the task or operation is done.
4. *Forms, records, documents, books, or files*: These are miscellaneous items that sometimes appear as single items and sometimes as part of other documentation.

Each organization defines the need for documentation, and these layers may indeed overlap from company to company. For these four levels of documentation to be fully

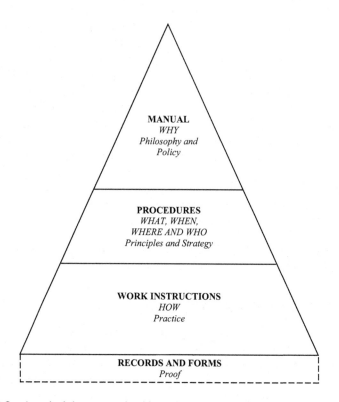

FIGURE 4.3   A typical documentation hierarchy.

effective, a cross-reference system must be developed to show the interrelationships between layers. These interrelationships within a total documentation system are shown in Figure 4.3.

The ISO certifiable standards require approval for all classes of documentation. The approval should be at an appropriate and applicable level within the organization. Generally, the higher the level of documentation, the higher the approval authority. A rule of thumb – from the author's experience – is that documentation approval should be from two levels higher than the level seeking the approval. When the complexity of organization or product/service is high, the approval process may be assigned to more than one person. Otherwise, one person should be responsible for granting approval.

Another requirement of the ISO standard is a control mechanism to ensure that only pertinent documents are used. Pertinence in this case applies to both the relevance of the subject matter of the document to the work at hand and the currency of the contents. This control entails issuing the documents needed at the correct revision level wherever and whenever they are required and ensuring that they cannot be inadvertently replaced by documents relating to another subject or to the correct subject but at a different issue. Therefore, all unnecessary documentation must be removed from the workplace without delay, and all revisions must be correct and current.

TABLE 4.1
A Typical Document Control Form

Copy Number	Version Number	Version Date	Copy Holder

When there is a need to copy documents – perhaps for marketing or public relations purposes – control of the copies must be proportional to the risk and their use. A unique document format, special paper color, or the incorporation of a colored mark in all legitimately issued documents all make satisfactory controls.

Alteration to documents must be controlled. To prevent irrelevant changes, the standard requires that the document shall be reviewed and approved by the original reviewers or by other authorized people thoroughly conversant with the background information on which the original documents were based. In order to ensure that the correct documents may be issued and withdrawn, a control mechanism must exist to identify documents, revision levels, and their holders. Table 4.1 shows a typical document control form.

THE QUALITY SYSTEM MANUAL

Even though the QM is not a mandatory document, we feel that it is an important document and it should be considered as part of the documentation bundle. Obviously, it is not necessary. However, the included information must be incorporated and published someplace.

So, the purpose of the quality manual is to describe at a high level of detail (30–60 pages in length) the organization's quality policy, vision, and quality system in order to

- Provide a reference point from which a reader can identify and locate specific procedures and operating instructions.
- Advise the organization's staff of the full range of standards and procedures and the interrelationships among them.
- Ensure common understanding of the way the firm does its business and how it regards quality.

The quality manual is not the full documentation of the quality system. The full documentation consists of the quality manual, procedures, work instructions, and records as shown in Figure 4.3. Because the quality manual is an overview and the first tier in the documentation, it is likely that individual task performers will not use it very frequently. In fact, the primary readership of the quality manual is the management, staff, and auditors (both internal and external).

Typical Contents of the Quality Manual

The quality manual should contain the following items:

- The organization's quality policy statement
- Proper authorization
- Issuing date
- The organization's principles and objectives
- The relevance of the specific ISO clauses to the organization
- A short description of the products/services
- A short description of the customers and suppliers
- Organizational structure
- Overview of the functions of the organization's executives
- Description (1–4 pages) of each functional area and its relationship to quality
- Cross-reference between functions, procedures, and work instructions
- Cross-reference between the organization's functions and the ISO requirements and other standards or customer conditions
- Distribution list
- Document change and control procedures and responsibilities
- Copyright statement (optional).

The scope of a manual should be site specific because certification is site specific. Therefore, a corporate quality manual may summarize the entire quality system, but when there are multiple divisions, each should have its own quality manual. Sometimes, when there is a need for an abridged version to indoctrinate new employees or to brief management the quality manual is printed in a pamphlet format. When literacy is a problem in the organization, the pamphlet format allows employees to carry the document in a pocket and review it as needed.

When confidentiality is a concern, the quality manual may be published as a quality policy (same as the quality manual but with the sensitive data removed) and distributed to customers and suppliers without divulging confidential information. Such a manual may be defined as a noncontrolled document. The overall structure and content of a typical format for the quality manual is presented in a series of figures. Figure 4.4 identifies the parts (components) of the QM. Figure 4.5 identifies the items on the cover (title) page. Figure 4.6 identifies the components of the table of contents. Figure 4.7 identifies the components of the QM body.

Finally, an optional introduction may be used to summarize the contents of the quality manual. It may cover special vocabulary and information to help the reader and the auditor understand the company and its products or services better. By understanding the organization better, the evaluation will be more meaningful.

Specific Requirements of a Typical Quality Manual

Since the QM is not a mandatory document, we will not review the individual clauses. However, some observations are in order. The importance and high visibility of the QM – both internally and externally – warrant a detailed discussion of each

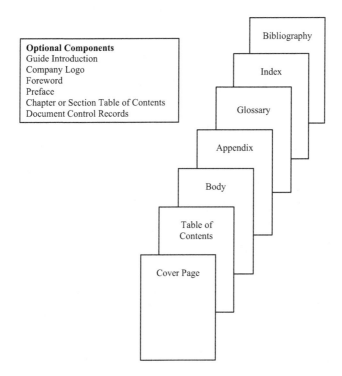

FIGURE 4.4 Typical components of a quality manual.

Document Reference:
Revision Number:
Date Issued:
Supersedes:
Section:

Company Name
Manual Title and ID

FIGURE 4.5 Quality manual cover page.

Document Reference:
Revision Number:
Date Issued:
Section:
Page_____of_____

Table of contents

| Section | Section | Revision | ISO | Page |
| Number | Title | Number | Standard | |

FIGURE 4.6 QM table of contents.

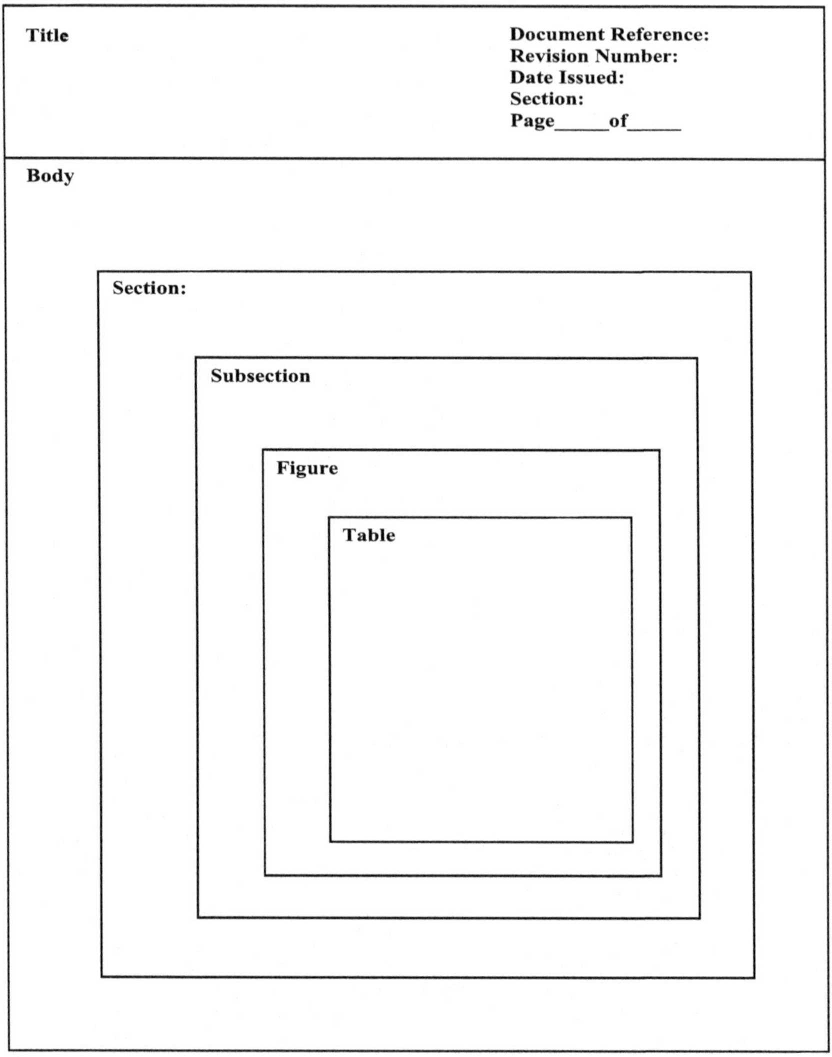

```
┌─────────────────────────────────────────────────────────────────┐
│  Title                               Document Reference:          │
│                                      Revision Number:             │
│                                      Date Issued:                 │
│                                      Section:                     │
│                                      Page_____of_____             │
├─────────────────────────────────────────────────────────────────┤
│  Body                                                             │
│      ┌──────────────────────────────────────────────────┐        │
│      │  Section:                                         │        │
│      │     ┌──────────────────────────────────────┐      │        │
│      │     │  Subsection                           │      │        │
│      │     │    ┌──────────────────────────┐       │      │        │
│      │     │    │  Figure                   │       │      │        │
│      │     │    │   ┌───────────────────┐    │       │      │        │
│      │     │    │   │  Table             │    │       │      │        │
│      │     │    │   │                    │    │       │      │        │
│      │     │    │   └───────────────────┘    │       │      │        │
│      │     │    └──────────────────────────┘       │      │        │
│      │     └──────────────────────────────────────┘      │        │
│      └──────────────────────────────────────────────────┘        │
└─────────────────────────────────────────────────────────────────┘
```

FIGURE 4.7   QM body components.

individual element. The inclusion of each clause should take into account the requirements of ISO 10013, which is the official guide for writing a Quality Manual. The requirements should not be very detailed but rather generic in nature, and they may not apply to all organizations. What is important is the notion that the QM defines the vision of the organization, identifies the appropriate documentation, and references the appropriate tools for such documentation. *The proof, however, of their existence will be established during the actual audit.*

Procedures

A pictorial overview of a typical documentation scheme can be seen in Figures 4.1 and 4.2.

TABLE 4.2
The Four Tiers of Documentation

	Tier 1	State policy and objectives for
	Quality manual	each of the pertinent elements of
	Company product	the standard (generally the ISO
		elements)
Say what you do, do	**Tier 2**	Global objectives: Demonstrate
what you say	Department procedures (focus on	that you do it
	Who, What, When, and Where)	
What is it?	**Tier 3**	Details of how to do specific tasks
Where does it apply?	Work instructions	
When does it apply?	How do you do it?	
Who is responsible?	Equipment/instructions	
	Tier 4	Where is the information recorded
	Forms, records, books, logs, files	or filed?

The quality manual identifies the organization's policies and functions, and places them in a context that describes the relationships between them. A procedure identifies the procedural steps for a function. Depending on the scope and complexity of the function(s) (here may be one or more procedure manual(s)). Some examples of procedure manuals are procurement manuals, operations manuals, and manufacturing and engineering manuals. Each of these manuals relates to the procedures of a particular department. If the documentation is in an electronic form rather than on paper, it may not be necessary to divide it into separate manuals (see Figure 4.2). However, access to the procedures may be restricted on a need-to-know basis. A typical four tier documentation scheme is shown in Figure 4.3 and Table 4.2.

Instructions

The terms operating instructions, instructions, and SOPs are used interchangeably. They all describe details of specific steps of higher-order (less detailed) procedures. For example, a machine set-up may be a step in the total calibration procedure. The SOPs for the set-up is the specific description (the process, the method) for this particular kind of set-up. As a general rule, instructions are at a lower level of detail than both the Quality Manual and the procedures. Instructions describe a step-by-step approach to guide an operator through a task.

If the overall procedure is short and simple, the operating instructions may be part of the procedure manual or they may be bound separately. Because instructions are detailed and specific, they may be posted in the work station, provided to the operators as job aids, or made available on demand through electronic devices. No matter how the instructions are distributed, all instructions must be available to the operators performing the task or the process and they all must be current.

Contents of Procedures

Although there is no definite way to identify specific content, there are some generic ways to build a procedure. The following questions may help in structuring a procedure – see Table 4.3.

TABLE 4.3

A Generic Approach to Structure a Procedure

Purpose, objective	Why Is This procedure performed?
Inputs	What materials come to the process? What input need to be measured or evaluated?
Outputs	What specific outcomes are expected? Are there any interim and/or product measurements?
Acceptance criteria	How are the tests, standards, or tolerance limits selected?
Procedure steps	How is the process performed?
Responsibility	Who is responsible for initiating, performing, and monitoring the process?
Audit requirements	How is the frequency of the audit determined? Who performs the audit? What happens to the results?
Resources	What resources (manpower, machine, method, material, measurement, and environment – 5Ms and E) are required to perform the procedure or instruction?
Training requirements	What training courses, seminars, workshops, etc. are required for the operators of the particular task?
Approval authority	Who approves and authorizes the procedure or instruction?

Although all of the elements in Table 4.3 are not mandatory in every procedure, the organization must define the applicability of the content and as a consequence incorporate the appropriate material. The level of detail for each procedure is unique to itself and the organization, and there is no special or predetermined length for any procedure. However, to ensure maximum applicability and appropriateness of the individual task, the operator must be included in defining the task. The following guidelines may help in procedure writing:

- Provide an overview that includes the procedure's purpose.
- Identify all materials, tools, and ingredients for the procedure.
- Identify both prerequisite and requisite skills.
- Provide an estimated time to complete the procedure.
- Do not assume. Do not skip obvious steps.
- Follow the correct order (logical, hierarchical, procedural, etc.). If order is not important, say so.
- Each step should be simple and self-contained.
- Use a hierarchical structure whenever possible. It is very easy to develop.
- Start with the most important step and keep developing them until the size and complexity of the steps give you the desired results.
- Try to start each instruction with an active (direct) verb (e.g., place, drill, position) or with the prerequisite, if there is one.

- Number each of the steps and use the numbering system consistently throughout the documentation (e.g., the numerical system of the standard itself or numerical [#.#.#]; roman [I., II.]; alphabetical [A., B.]; or combination [I. A. 1.a.i.]).
- Always verify the procedure. Test it against current knowledge. Do not assume the procedure is correct because you just finished writing it.

Formats for Procedure Writing

In any system, one may select from a choice of formats available. Some typical and common ones are

- *Paragraph form (not recommended)*: Hard to write, hard to maintain, and hard to read and use.
- *Numbered instructions*: Used when the sequence or hierarchy is important and must be identified.
- *Playscript*: Used with low-literacy employees. Performers are identified with the steps in the procedure they perform.
- *Bullets and hyphens*: Similar to numbered instructions but the sequence is not important.
- *Process flow chart*: Used with long and semi complex processes. Traditional symbols identify each task.
- *Parallel form*: An annotated process flow chart. Highly recommended for all processes. Instructions are keyed to the process chart.
- *Interactive dialogue*: Similar to the numbered instruction format but with the addition of responses from a machine, usually a computer. Expensive to maintain.

In addition to the specific content, each page of the procedure should have the following information:

- The name of the procedure
- The name of the organization
- The authorizer
- The page within the procedure (page #, or the preferred way: page: _____ of _____)
- The effective date
- The revision date

This information may be split between a header and/or a footer within the page as in Table 4.4. However, it is recommended that all the information appear as a header on all pages. Consistency is important.

In some cases, job aids are necessary. They are specific cards that usually contain excerpts from the quality manual. They give operators easy access to specific instructions, overviews, and reminders. Typically, the cards include highlights or outlines of procedures or guidelines such as tolerance limits for specific tests. The nature of the work process dictates the content and function of job aids.

TABLE 4.4
Procedure Header and Footer Example

Title of procedure:	Effective date:
Authorized by:	Revision:
Preparer:	Page _____ of _____
Contents:	
Date issued:	Company/department name:
Date revised:	Reference number:

THE PROCEDURE AND INSTRUCTION DEVELOPMENT PROCESS

The focus of documentation is to describe accurately the current state of the organization and to demonstrate conformance and/or compliance to the appropriate standard. The documentation – quality manual, procedures, and instructions – must demonstrate to the customer and auditor that conform and/or comply to specifications that the customer requires and compliance as to what the standards – especially government regulations – require in an objective way so that there is no ambiguity in translating the intent or effect of the result. To make this effective and easy to implement, a five-step approach is recommended:

1. *Determine requirements*: What does the standard require? Is the standard appropriate? What do we have to do as an organization to meet the requirement?
2. *Perform a need (gap) analysis*: What is the gap between the existing documentation and the requirements of the standards? Usually, this is performed with an *internal audit*.
3. *Define the documentation effort*: What tasks must be performed to add new documentation and revise existing documentation to comply with (the standard? Who will perform them? When? How?)
4. *Develop and refine the documentation*: Define and document the roles and responsibility of the appropriate people, organizational structure, appropriate standards, and procedures. Write the documentation, and verify that it is accurate and usable. Publish the documentation. Distribute the documentation to the people who will use it.
5. *Control the documentation* throughout its life in accordance with the procedures established for document control in the quality manual. (Update, as necessary.)

EFFORT AND COST

The level of effort required to develop quality documentation depends on the current state of the organization's quality standards and procedures. A given organization may have adequate and documented standards, or have inadequate standards. Standardization of the formats as much as possible throughout the writing and editing process will make the job much easier. Of course, other ways to develop documentation are through teams or through the business structure of the organization. If the team approach for documentation development is used, the team should be made up of people who are multifunctional and multidiscipline as well as having knowledge of the process or task that they will be writing about. So, the team approach should be focused on individuals who

- Are knowledgeable about the process.
- Have been trained appropriately in the areas of the standards, group dynamics, and problem-solving techniques.
- Are considered leaders by their co-workers.
- Have the authority and the time to do the work.

Once the team is in place, the development of the quality system can begin. At least three areas of investigation and analysis are important.

1. Where are the quality events performed? (Quality events can be defined as reviews, inspections, tests, and audits. These events may be internal or external to the organization.)
2. What standards will be applied?
3. What specific procedure steps will be performed throughout the process and at the quality events?

The authority to investigate and/or analyze these areas must be defined early on and must be visible to everyone. A good general rule to follow for appropriate approval in any documentation situation is the two to three levels above the level at which the documentation is used. If multiple utilization is expected, multiple approvals may be required.

Having visible authority is not the only parameter that will define the success of the documentation. In fact, visible authority may sometimes inhibit success. What is really necessary is appropriate knowledge of the quality events. In a typical internal manufacturing setting, quality events occur

- Upon receipt of material.
- At key points in the manufacturing process at which the identification of a discrepancy (nonconformance) can avoid unnecessary expense or effort farther down the line.
- At the completion of the manufacturing process.
- At the time of shipment.

In a typical external manufacturing setting (with a supplier certification program), inspections and product verification to a set standard can be performed by the supplier. The purpose of reviews, inspections, and tests is to identify deficiencies in the process or the product in order to avoid unnecessary effort and to ensure that the end product is acceptable. Furthermore, they provide one more input for continual improvement.

Audits of quality events monitor conformance and/or compliance with standards and procedures. A quality event is the quality attained for a specific task as defined by the requirement(s). Audits on quality events may be conducted at regular intervals or be completely unannounced. While reviews, inspections, tests, and audits are very easy to develop and document, establishing design standards is a complex effort, often performed over many years as the product(s) evolve. In any case, design standards are usually identified through customer requisition, product planning, or as a result of product evaluation and improvement.

Finally, the contribution of the team may be affected by the administration procedures that quite often define the following:

1. How the organization documents standards, procedures, and quality events.
2. How the organization addresses the verification of compliance.
3. How the organization handles the variance process. (Variance here is any deviation from a standard. In the auditing language, a deviation – no matter how small or large – is referred to as nonconformance. If the audit is to confirm compliance, then the deviation is referred to as noncompliance.)

The second approach to documentation is the approach of the business structure. Because different organizations have different ways of handling organizational objectives, the ISO standards, for example, deal with the functions that are critical to product and service quality. As a consequence, the standards exclude the executive, sales and marketing, and accounting functions from specific responsibility. However, these functions make a major contribution to an organization, and without their support, the certification will not take place. For example, without executive commitment, how can the organization fulfill the ISO 9001 clause 4.1 or the IATF 16949 clauses 4 and 5? Without the contribution of sales and marketing, how can the organization define customer requirements and corrective action? Without the active participation of accounting, how can the organization calculate the figures for nonconformance, rejects, rework, warranty, *etc.*?

The business structure in some organizations may dictate that the documentation be based on departmental control, central control, or a business unit. In any case, what is important is the participation of the entire organization and everyone will contribute if success is to be the end result.

The cost for such an undertaking depends upon whether the organization has a quality system in place. The cost is minimal if the organization already has a workable quality system. On the other hand, the cost can be very high if the organization starts without a quality system. In the author's experience, the average cost for documenting an organization of up to 200 employees is about $150,000.

QUALITY PLANS

The ISO 8402(A3), clause 2.5.3, page 4 defines a quality plan as a document setting out the specific quality practices, resources, and activities relevant to a particular product process, service, contract, or project. However, another way of describing the quality plan is to define it as the entity that contains the standards and procedures for special situations. More often than not, these standards and procedures supersede the normal ones for the specific situation.

Sometimes, the term *quality plan* is used interchangeably with the terms *project* or *control plan*. However, because of its special orientation, the quality plan more often than not is required to cover instructions that are unique to a process, product, or customer. If the uniqueness is of contractual nature, then the quality plan may be a separate document. Nevertheless, in most organizations the quality plan may be part of the work order or print or an addendum to the specific plan.

One of the most often asked questions in reference to quality plans is: What situation warrants a quality plan? To answer this question, we must first look at the customer requirement and accordingly provide for the specificity to the standards and procedures. Second, care should be taken to correlate the quality plan clearly with the specific situation to which it applies and to provide a means for signaling the situation to the task performers.

This second requirement is very powerful, for it allows the task performer to demonstrate empowerment at the job location as well as demonstrate the quality commitment to prevention as defined by the organization. For this demonstration (validation) to take place, the performers must come in direct contact with the project, product, client, etc. so that they can identify as closely as possible the work being performed and then choose the appropriate quality procedure. To facilitate this process, posters and job aids may be used to remind the performers of the special handling procedures.

If the quality plans are extensive, some organizations may use a master index (Master Control Document) to assure consistency and properly identify all special procedures with the appropriate jobs on the work orders. To ensure consistency in even the most complex procedures, it is not uncommon to establish the basic (generic) procedure for the plan and then add each requirement as needed. Examples for this situation are Failure Mode and Effect Analysis (FMEAs), G8Ds (Ford's Problem-Solving Methodology), and Control Plans.

To demonstrate this generic procedure, let us look at a chemical organization. The organization has many processes with many individual (unique) products. One way to address the quality plan is to identify each product with its own idiosyncrasies. Another and more efficient way is to define the generic process and then refer to the products as they are produced. The generic procedure is as follows.

Procedure for Process Y

1. Get or receive materials.
2. Place material in ladle.
   See Table Y.1 for specific measurements.
3. Mix well until predetermined viscosity is reached.
   Table Y.1 Measurements
   Standard measurements
   X lbs. of A
   Y lbs. of B
       Measurements for client CC; Contract 1299
       S lbs. of A

T lbs. of B
Measurement for client SP; Contract 15668
U lbs. of B
etc.

Typical information included in a quality plan per ISO 9004:2018 clause 5.3.3 follows:

- The quality policy statement.
- Description of the particular situation.
- The objectives to be attained.
- Acknowledgment and approval from the responsible authority(ies).
- Description of the products, services, customers, and suppliers.
- Organizational structure unique to the situation.
- Overview of the procedure or process.
- Roles and responsibilities unique to the situation.
- If testing or inspection is performed, the criteria and testing, review, or auditing should be identified.
- Description of each functional area's procedure.
- Cross-reference matrix between functions, procedures, and operating procedures.
- Distribution list.
- Document change and control procedures and responsibilities.

Even though the list seems to be quite extensive and inclusive for specific requirements, it must be emphasized that these are only guidelines. The specific quality plan may not include every item, but in some cases, it may have more items, depending on the requirements of a particular customer. (In the automotive industry, the control plan is standardized through the Automotive Industry Action Group (AIAG) with 26 required items. More may be added, but none may be eliminated.)

QUALITY RECORDS

ISO 9001 clause 4 addresses the issue of quality records by stating the supplier shall establish and maintain documented procedures for identification, collection, indexing, access, filing, storage, maintenance, and disposition of quality records.

How does one really define a quality record? A quality record is the result of the performance of a task, audit, inspection, test, and/or review. Generally, a quality record provides the basis for analysis of the work in process and is used to determine corrective action and improvement to the process and to provide the basis for comparison (compliance) to standards or auditing certification.

Quality records may be generic or very specific to a given organization. However, all quality records fall into the following categories:

1. *Process control results*: They measure various aspects of both the products produced and the process itself. Based on these records, adjustments, improvements, and automation are introduced wherever applicable and appropriate. Important characteristics of all process control results are the issues of

a. *Retention of records*: How long do we keep the records?
b. *Sampling size*: How much of a sample is adequate?
c. *Frequency*: How often do we sample?
d. Appropriate statistical analysis? What statistical analysis do we use? Why?

2. *Inspection and test results*: Both are means by which humans or automated equipment use preestablished acceptance criteria to evaluate in-process products, end products, components, materials, processes, or services. Generally, the data provide a means to prove that tests were performed, to evaluate the effectiveness of the process, and to identify problems. Inspection results, on the other hand, verify a set quality expectation. Important characteristics of all inspections and tests are

a. *Date and time of test or inspection*: When was the test or inspection performed?
b. *Test or inspection identification*: What is the specific name of the test? What is the specific approach of inspection (*e.g.*, stratified, sequential, or randomly sampled)?
c. *Test criteria*: What is or is not acceptable? How are they determined?
d. *Inspector or tester*: Who performed the test or inspection?
e. *Results*: What are the results? How do the results compare with the test criteria?
f. *Follow-up*: Is a follow-up required? Who is responsible for the follow-up? How much time is available for the follow-up? Is corrective action required? Is improvement required? Does a deficiency need to be tracked and fixed? Is there a provision for prevention?

3. *Audit results*: They are the documentation (records) of the audit itself. They may be formal (form) or informal (memo). Specifically, they should include the initial response, the findings, the resolution, and the expected action based on the findings (follow-up). Important characteristics of all audit results are

a. *Date and time of audit*: When was the audit performed?
b. *Process audited*: What was audited?
c. *Auditor's name and organization*: Who is the organization being audited? Who performed the audit?
d. *Responsible management organization*: Who authorized this audit? Who is receiving the results?
e. *Discrepancies and outstanding findings*: What discrepancies (noncompliances or nonconformances) were found? What outstanding items were found? Each finding should have the following information:
    i. Identification item.
    ii. Standard associated with the particular finding. Time frame to correct.
    iii. Criticality of noncompliance or nonconformance.
f. *Overall result*: Did the organization pass the audit? The result is always pass/fail and never a numerical score. Some audits for specific customers require a numerical value, for example, the Q1 3rd edition of Ford Motor Company.

g. *Audit report distribution*: Who receives a copy of the audit?

h. *Checklist*: Audit checklists are part of the audit documentation. They show the audit criteria or the subject areas of the audit. Has a checklist been developed? Are all areas represented in the checklist?

An audit response is part of the quality record, and its content should include the following:

- All responsible parties
- Reference to the audit report.
- Acknowledgment of deficiencies reported and acceptance.
- Acknowledgment of all issues not recognized as such and they are not accepted
- Description of corrective and prevention action(s)
- Request for re-audit or review (if needed)
- Time table for removing deficiencies
- Date of corrective action.

4. *Training records*: They are all the records associated with the employee's training. They may be found either in each employee's department or in the Human Resources Department or both.

5. *Product and process review*: Product reviews ensure that interim and end products comply with the standards and customer requirements. Process reviews ensure that ongoing (formative) performance evaluation is effective. It is very important that process reviews be performed by the performers (operators) of the task on a continual basis, and they generally follow the audit format. Their focus is process improvement rather than deficiency identification.

6. *Logs and service calls*: Logs are usually informal documents listing particular discrepancies, review results, and events. Service calls, on the other hand, are documents relating to customer (internal or external) communication in reference to a particular complaint, discrepancy, etc.

To identify the quality records in a given organization, one must keep thinking of the appropriateness and applicability of the record to the organization. The ISO 9004 states, "the system should require sufficient records be maintained to demonstrate achievement of the required quality and verify effective operation of the quality management system." The following items are some of the types of quality records that require control:

- Inspection reports
- Test data, plans, and acceptance criteria
- Qualification reports
- Validation reports
- Audit reports
- Material review reports
- Calibration data
- Quality cost reports.

It is important to recognize that these records mentioned in the ISO 9004 are minimum events that occur across the organization's process. As a consequence, additional reports may be required to satisfy the overall quality management system.

Two other aspects of the control of quality records are the issues of: First, the issue of retrieval. ISO 9001 clause 7.5.2–7.5.3.2.1: Where agreed contractually, quality records shall be made available for evaluation and adequately protected by the customer or the customer's representative for an agreed period – clause 7.5.3.2.1. "In essence, the standard requires that quality records should be retrievable by some criteria. Secondly, on the issue of retention, the same clauses mentioned call for the quality system that should provide a method for defining retention times, removing and/or disposing of documentation when that documentation has become outdated." In addition, they mention that all documentation should be legible, dated, clean, readily identifiable, retrievable, and maintained in facilities that provide a suitable environment to minimize deterioration or damage and to prevent loss.

One can see quite clearly that the standard is explicit as to the need for retention; however, it does not provide specifics. The specifics of document retention need to be defined by the organization and the customer. Generally, for most companies the retention period is between 5 and 7 years or 1 year after production of the part. (In some industries, the retention is from *cradle to grave* such as in nuclear industry.) It is the organization's responsibility to retain the quality records intact and away from all hazards.

DOCUMENTS FOR DESIGN CONTROL

In some organizations, design is an integral part of the business. As such, specific documentation must exist to satisfy the several clauses in Section 8.0 of the ISO 9001 which state that the supplier shall establish and maintain documented procedures to control and verify the design of the product in order to ensure that the specified requirements are met. The procedures to ensure that requirements are met are essentially the project planning and control process procedures.

Projects are efforts to achieve an objective in a finite lime and within finite budget. When the objective is achieved, the project is over. With an ongoing activity, the process is perpetuated. In projects, one must always be cognizant that the quality assurance events are checkpoint reviews (milestones), which are mapped to the major phases of the project.

The objective of the reviews is to validate that the work performed to date is acceptable. The reviews are performed throughout the project. On the other hand, the objective of the checkpoint review is to validate the entire results of the project phase and to re-evaluate the go/no go decision. Checkpoint reviews are special reviews. The focus of both reviews and checkpoint reviews is to minimize the impact of errors by uncovering them early in the process, to avoid unnecessary effort by readdressing the go/no go decisions, and to uncover opportunities for process improvement. (A typical milestone review is the Advanced Product Quality Planning – APQP.)

Each project may be totally unique and unrelated to anything else. Therefore, is it possible to really focus on a standard documentation? The answer is a definite yes. The reason for such a positive response is that in all projects, the *project life cycle* is a standard set of steps that can be applied to any kind of project. In fact, the following characteristics – also shown in Figure 1.1 (project life cycle) – are major tasks that should be the focus in any documentation endeavor.

Project initiation or feasibility study: Typical documentation may be input requirements, cost-benefit analysis, design, estimates, tasks, roles and responsibilities, and statement of feasibility.

- *Requirements definition*: Typical documentation may be review of requirements, analysis, resolution, regulatory requirements, fitness for purpose, and safety requirements.
- *Product design*: Typical documentation may be detailed specifications, design reviews, and reliability.
- *Prototype development*: Typical documentation may be request for changes, design results, design reviews, follow-up requests, and checkpoint reviews. In the prototype development stage, we are concerned with detailed design, development, and integration and testing.
- *Implementation*: Typical documentation may be tests, validation, reliability, evaluations, acceptance criteria, and cause identification.
- *Post-project evaluation*: Typical documentation may be specific verification techniques, formal reports of deficiencies identified during implementation, and corrective act ion reports.

To guide us in the documentation process of a design with some specificity, the ISO 9001 clause 8.3 identifies some requirements. They are not exhaustive nor are they intended to fulfill every requirement of a design situation. A selected summary of the requirements is as follows:

- 8.3.2 *Design and development planning*: Create a project plan that identifies the activities (tasks) to be performed, the responsible performers, and their organizational and technical interfaces
- 8.3.3 *Design input*: Identify, clarify, accept, and document the business requirements that are the inputs to the design process and the ultimate criteria for success of the project.
- 8.3.5 *Design output*: Document the detailed specifications, calculations, etc. that represent the design, and
  - Satisfy input requirements
  - Identify acceptance criteria (for the product)
  - Conform to regulatory requirements
  - Identify safety characteristics of the design.
- 8.3.4(b) *Design review*: Records for the documented reviews of the design, which will include all planned activities and representation involved during the design and development, as well as design input stages
- 8.3.4 (c, e) 4.4.7 *Design verification*: Plan, staff, and document the procedure for verifying that the design output meets the design input requirements by means of control measures such as
  - Design reviews
  - Tests and demonstrations
  - Alternative calculations
  - Comparing the new design with similar proven designs.

- 8.3.4 (d); 8.3.4.2 *Design validation*. Design appropriate validation procedures so that the product conforms to defined requirements.
- 8.3.6 *Design changes*: Identify, document, review, and approve all changes or modifications to requirements and design

DOCUMENT AND DATA CONTROL

One of the most vulnerable areas in the ISO standards and customer-specific requirements is the area of document control. Here, we will emphasize the ISO requirements because they are the very basic requirements for all other instances where they appear. So, we see

- 7.5.2 Document approval and issue
  - Establish and maintain procedures to control all quality documentation.
  - Ensure that documentation is reviewed and approved by authorized personnel prior to issue.
  - Ensure that only the current version of a document is available where and when needed.
- 7.5.3 Document changes/modifications
  - Review and approve changes.
  - Identify the nature of the change.
  - Maintain a control procedure to identify the current revision of a document and its location and ownership to preclude the use of nonapplicable or outdated documents.
  - Reissue documents after a "practical" number of changes have been made or when documents are illegible, messy, etc.

ISO 9004 goes even further in addressing the subject of document control: Pertinent sub-contractor documentation should be included. All documentation should be legible, dated, clean, readily identifiable, retrievable, and maintained in facilities that provide a suitable environment. Data may be hard copy or stored in a computer.

Often the question arises as to what constitutes document control. The answer is not an easy one. It depends on the organization and the documents needed for verification of a particular task. Generally, documents and records may sound alike, but there is a big difference between the two. Documents are created by planning what needs to be done and records are created when something is done and record the event. Documents can be revised and change, whereas records don't (must not) change. To help define document control, the ISO 9004 which provides the following guidelines:

- Product documentation (drawings, specifications, blueprints)
- Inspection instructions
- Procedures (tests, work, operational, quality assurance)
- Operations sheets (process documents)
- Quality manual.

Although these are very general guidelines, document control fundamentally addresses two issues: the document administration and the document change process. On the first issue, document administration, document control consists of the following:

- Obtaining original content or changes.
- Technical editing.
- Ensuring that content is reviewed and approved.
- Publishing the documented material.
- Keeping track of the location and ownership of the documents.
- Ensuring the security of the document.
- Distributing the changes promptly and with appropriate removal instructions.
- Denoting changes according to the documentation standard.
- Replacing documents after a practical number or changes have been made.

The responsibility for the review and approval of the contents should be with the people who will have to comply with the contents. After all, *what is being documented is the way work is being performed, not the work that is going to be performed.*

On the second issue, document change process, document control consists of the following:

- Identifying of the required change(s).
- Writing the change(s).
- Approving the change(s).
- Distributing of the change.
- Including the "new" documentation.
- Removing the "old" documentation.

Distribution control is one of the most critical functions of document control. It is necessary to ensure that documents are distributed to the appropriate people or locations and that changes to the documents are distributed so that the documents can be kept up to date. Anyone with knowledge of the performance, task, process, *etc.* may propose the changes. On the other hand, only with the person – or the equivalent of his or her position – who authorized the original can approve the change. As a general rule, approvals are performed by a person who is at least two levels higher than the seeker of the approval.

When documents are distributed in a hard copy form, the document must be maintained in a master file by the document administrator (see Table 4.1). Typical contents of the master list are

- Title
- Copy number
- Version number
- Version date
- Holder of the copy
- Date of last revision (optional).

Not all documents are controlled documents. Some documents are uncontrolled because they are not used in the actual work process. When that is the case, the copies must be clearly labeled to avoid inadvertent use (common identification is a stamped message that identifies them as *guidelines*. These guidelines ARE NOT audited. Also, it may be a common identification as a stamp notifying the reader of the status of the document. The stamp may read *controlled* or *uncontrolled document*; *classified* or *unclassified* document; or any other statement reflecting the status of the particular item. Sometimes, the differentiation may be through the use of a different color ink or a completely separate stamp.

Document administration is an ongoing process in the organization and is clearly stated in the quality manual or in some other document. The document change process, on the other hand, is a process that is defined in the quality manual and that actuates itself when needed. The frequency of change, for example, may be defined as a result of a change in the process due to improvements or to changes in standards, policies, roles, or responsibilities.

Very frequent changes in procedures will cause a great deal of administrative and control work for those responsible for documentation control. Because of this lengthy work effort, it is recommended that procedures be constructed in generic terms and that the individual quality plans or instructions be constructed in very detailed terms. Frequent changes may be the result of performing custom work for individual customers; the changes are unavoidable, however, welcome the work.

When frequent changes are common, and not profitable to the organization at large, a review of the structure and objective(s) of the documentation may correct the situation. It turns out that the most frequent causes for such changes are lack of procedure analysis and development. A good tool for this analysis is the *process flow chart*.

A document should be reissued when a single change or a number of changes have been made, either over a period of time or at one time, that could be misinterpreted or confuse the reader. Another reason to reissue a document is the physical condition of the document itself. In the case of electronic distribution, security is also an issue. The document control administrator is responsible for updating and maintaining the functionality of the documents. Restrictions should be placed on printing and modifying all the documents in the electronic distribution system. Even though the electronic system provides an easier and more effective system for changes and control, it should be remembered that once a quality document has been printed, hard copy controls must apply.

In the electronic system, document control usually consists of the availability of access at the work stations where the documents are used and of security features that are part of the electronic distribution system such as passwords or special access codes.

PRODUCTION ISSUES IN THE DOCUMENTATION PROCESS

The appearance of the quality documentation is an important aspect of the documentation effort. Although the ISO/IATF standards do not require any particular level of aesthetic presentation, attractive presentations enhance both the readability and usefulness of the documentation. Production issues include layout, fonts, position, spacing, underlining, bolding, use of footers and headers, references, indexing, related documentation, tables, figures, and other pertinent information. The specific use and appearance of any of these items depend on the organization and the requirements of the ISO/IATF standards and/or the customer.

A word processor and desktop publishing and graphic software should be used to facilitate production of documentation. The use of good software will save a tremendous amount of time in revising and updating the information; 8½″ by 11″ paper (in the USA, A4 size in Europe) is usually used for presenting paper documents, and online documents may be distributed via a network, disks for standalone PCs, or a mainframe computer. The distribution is a matter of personal preference. Both methods require maintenance of the documentation, and the specificity of the maintenance should he discussed under document control. The components of the documentation follow a standard sequence, such as

- Cover page
- Table of contents
- Foreword (optional)
- Preface (optional)
- Index or matrix
- Body (may be divided into chapters, sections, and table of contents)
- Glossary (optional)
- Appendixes (optional)
- Bibliography (optional).

A typical cover sheet (see Figure 4.5) will include the following items:

- Company name
- Manual title
- Identification signature
  - Prepared by
  - Revision date
  - Copy number
  - Authorization signature.

A typical table of contents sheet will include the following items, but not necessarily in this same order (see Figure 4.6.):

- Title
- Table of contents
- Section title
- Section number
- Page
- Authorization
- Last revision (optional).

The Body of the Documentation Should Contain the Following Items

Document title (section explanation). The appropriate detail of this section is of importance so that the reader gains the information without turning the page or the screen. The explanation is usually given in a form of a flow chart; however, in some cases, the prose form is used. In the case of the computer scrolling, it can be used; however, limit its usage to only when it's absolutely necessary. Its usage disrupts the readers' concentration and therefore IS NOT recommended. Typical items are

- Date
- Revision date
- Page number
- Company logo.

Figures, tables, and forms (optional). If figures, tables, or forms are used, whether in a hard copy or electronic media, special care should be taken not to split these elements between pages or screens – unless absolutely necessary.

Basic Issues of Technical Writing

A final consideration in writing quality documentation is good technical writing. It is imperative to write the documentation with some form of standardization and some consistency. The issue of consistency is very important, since the documentation will undoubtedly have more than one author.

To focus on the mechanics of writing, we recommend a writer's guide, or perhaps fill-in-the-blanks outlines or word processor documents to reduce the writing effort and to maintain maximum consistency and standardization.

In writing the actual documentation, we recommend the following:

- Follow the *KISS* (Keep It Simple and Sweet) principle.
- Write short and simple sentences; try to follow the rule of one sentence, one thought.
- Use the right words – avoid jargon as much as possible. Use the ISO 8402 – A3 as a guide to special words.
- Use active or direct verbs.
- Write short paragraphs.
- Use lists as much as possible, such as
  - Words
  - Phrases
  - Sentences
  - Pictograms.
- Use flow diagrams
- Use the right tone. Write to the reader's level of language and technical knowledge.

Edit, review, examine the readability, and proofread the material. Editing, reviewing examining, and proofreading may occur at any stage of the writing and should consider the following concerns:

Content: Is the content accurate and complete? Is it usable? Does the content reflect the present system as opposed to what should be? The reviewer must be familiar with the content; appropriate reviewers are the operators, supervisors, or managers.

Usability: Is the document written for the intended audience? Is the vocabulary appropriate for the audience? Is the structure of the document appropriate for its intended purpose? Are the size, volume, graphics, and flow of the documentation suitable, and useful? Because usability is of utmost importance, the reviewer should be the actual user of the document. In some cases, in fact, pilot tests are recommended to check the authenticity.

Compliance: Compliance must be based on the organization's conventions, standards, and specific organizational requirements. In the simplest terms, compliance and/or conformance are (should) be based on practicing what you say. So, are the headers, footers, font sizes, styles, content, etc., consistent with the requirements of the standards and the organization's conventions? Because this sort of editing requires special knowledge of standards and cultural conventions, we recommend that either a professional proofreader or a technical writer perform this task.

Grammar: Is the spelling correct? Is the structure of the writing correct? A word processor with a spelling checker and grammar checker is a very useful tool.

Production: Are the documents accurate, attractive, complete, etc.? This task should be performed by a technical writer or an editor.

A typical checklist for editing documentation may include the following areas:

- *Purpose of the document*: Is the purpose of the documentation clearly defined? Is this purpose constant throughout the chapter or section? Can this purpose be reasonably met by this document?
- *Audience*: Is the audience for the document clearly indicated? Is the document clearly oriented to this audience? Is the information organized appropriately for this audience? For this audience, does the document have the right level of technical information? Is an appropriate language level used? An appropriate level of detail?
- *Content*: Is the information complete? Has any irrelevant material been included by mistake? Are the concepts explained clearly? Is all information technically accurate? Have all procedures and operating instructions been tested? Are all terms defined clearly? Have all topics been given the appropriate emphasis?
- *Graphics and layout*: Are all necessary graphics included in the document? Is every graphic discussed or referred to in the text? Are the graphics reproduction quality? Is the document in standard format? Is there enough white space? Is the page layout user friendly?

- *Writing and formatting*: Have over-long sentences been shortened? Is the one sentence, one thought rule followed? Are sentences all the same structure? Should sentence length and grammar be more varied? Has the text been checked for errors in grammar, punctuation, and spelling? Can complex or fancy words be replaced by simpler words? Do transition words achieve flow? Are there helpful transitions from section to section? Are there headings or passive sentences that should be stronger? Is the style sincere, polite, and courteous? Has the reader's point of view been taken into consideration? Have others reviewed the document and provided feedback regarding its appropriateness, applicability, and usability? Does the reader understand each step? Does the reader know the best way to use the material, especially if the document pertains to SOPs or the operating instructions? Has the Fog index been used to measure readability? The Fog index is one of the simplest ways to determine the readability level of any document. The format and the calculation steps are

Step I
    Count the number of sentences _____
    Count the number of words _____
    Calculate average sentence length (words/sentences) _____ = _____ (rounded).

Step 2
    Count the words with three or more syllables (hard words) _____
    (Leave out the capitalized words, combinations of short and easy words [*e.g.*, grasshopper, bookkeeper, verbs ending in *ed* or *es*])
    Count the hard words _____
    Calculate the percentage of hard words [(Hard words/total words) × 100] = _____

Step 3
    Add average sentence length _____ + percentage of hard words _____ = Total _____

Step 4
    Multiply the total by 0.4 to get the readability level.

Note: A number equal to 9.0 indicates the material is for a ninth-grade-level reader. The number 14.2 indicates the material is for two plus years of college. The lower the number, the easier the reading.

PRODUCTION CONCERNS

Production and binding: Is the production style easy to use and update? Will the user be able to store and access the documentation easily? Is the cost acceptable? Will physical deterioration be an issue? Binding, reproduction, and distribution of the documentation are the final stages. In the case of binding, the following options are available:

1. *Three-ring binding*: Moderately expensive. Easy to add and remove pages, but the user might incorrectly arrange the pages. Highly recommended.
2. *Stapled binding*: Inexpensive. Looks very unprofessional. Difficult to add or remove pages.
3. *Spiral binding*: Fairly expensive. The document looks neat and lays flat, but material cannot be added or removed with ease.
4. *Folders or portfolios*: Inexpensive. Very unprofessional.
5. *Laminated pages*: Very expensive. Protects documentation in heavy use environments. New material cannot be removed or added unless the full page is replaced.
6. *Perfect binding*: The most expensive. Very professional, magazine quality. Material cannot be adjusted without reprinting entire documentation manual.
7. In today's world, most of the production issues – if not all of them – are computerized. In fact, in most organizations, items 1–6 – are considered noncontrol items – whereas item 7 is the control document.

In the case of reproduction, the following choices are available:

1. *Copying*: Inexpensive and fast for most of the applications of documentation.
2. *Offset printing*: Somewhat expensive, but if slick graphics and color are needed, this type of reproduction is recommended. Requires a commercial printer.

Finally, the document distribution will be based on the document control procedures of the quality manual.

KEY ITEMS FOR AUDITORS TO "WATCH FAR"

Key items that all auditors should be cognizant in their auditing practice of ISO 9001:2015 and IATF 16949:2016.

Document – Clause 7.5

- Information and the medium on which it is contained
- Information: As per 3.8.3, a meaningful data
- Data: As per 6.1.2.3, data is fact about and object (entity, item)
- Example: Record, specification, procedure document, drawings, report, and standards.

Record – Clause 6.1.2.3

- Document records of contingency plans, and verification/validation practices.

Documented Information – Clause 7.5.3

- Information required to be controlled and maintained by an organization and the medium on which it is contained.

MANDATORY DOCUMENTS

Determining the scope of the quality management system – 4.3

- The scope shall state the types of products and services covered.
- The scope shall provide justification for any requirement.
- The scope shall be available and be maintain as documented information.

Operation of organization's processes – 4.4.2a

- Documented information to support operation of organization's processes.

Quality Policy – 5.2.1–5.2.2

- Shall be available and be maintained as documented information.

Quality Objective – 6.2.1

- Shall maintain documented information on quality objective.
- Quality objective shall be measurable, monitored, and communicated.

Control of production and service provision – 8.5.1

- The documented information that define the characteristics of the products, the service, or the activity to be performed for which results to be achieved.

MANDATORY RECORDS

Operation of organization's processes for QMS (no such heading under this sub-clause) – 4.4.2b

- Retain the documented information.
- Information that has confidence that the processes are being carried out as planned.

Monitoring and measuring resources – 7.1.5.1

- Resources are suitable for the specific type of monitoring and measuring activity.
- Resources are maintained to ensure their continuing fitness.

Measurement traceability – 7.1.5.2

- Calibration to be traceable to international or national measurement standard.
- When measurement standard does not exist, the basis use for calibration or verification shall be retained as documented information.

Competence – 7.2

- Necessary competence of person shall be determined.
- Competence on basis of education, training, or experience.
- Action taken and evaluation of the effectiveness of the action taken
- Retain documented information as evidence of competence.

Operational Planning and Control – 8.1

- Documented information to have confidence that the processes carried out as planned and conformity of product and services as per requirement.

Review of the requirements for products and services – 8.2.3

- The organization shall ensure that it has ability to meet the requirement for products and services to be offered to customers
- Retain documented information of the review and any new requirement.

Design and Development Planning – 8.3.2

- Documented information needed to demonstrate that design and development requirements met.

Design and Development Inputs – 8.3.3

- Determine the requirements essential for the specific type of products and services to be designed and development, and retain documented information for these inputs.

Design and Development Controls – 8.3.4

- The organization shall apply controls to the design and development process; activities such as review, verification, and validation shall be carried out; and documented information shall be retained.

Design and Development output – 8.3.5

- Retain documented information for outputs.

Design and Development Changes – 8.3.6

- Retain documented information for design and development changes, results of review, authorization changes, and action taken to prevent adverse impacts on design and development.

Control of Externally Provided Processes, Products, and Services – 8.4

- Retain documented information for criteria of evaluation, selection, monitoring of performance, etc.

Control of Production and Service Provision – 8.5.1

- Documented information shall be available for the characteristics of the products to be produced, the service to be provided, or the activity to be performed and result to be achieved.

Identification and Traceability – 8.5.2

- The organization shall control the unique identification of the output when traceability is a requirement, and shall retain documented information.

Property Belonging to Customers or External Providers – 8.5.3

- Retain documented information for property of customer is lost, damaged, or found to be unsuitable for use and report to customer.

Control of Changes – 8.5.6

- Retain documented information related to changes.

Release of Products and Services – 8.6

- Retain documented information about acceptance criteria and traceability of person authorizing the release on the release of products and services.

Control of nonconforming outputs – 8.7.2

- Retain documented information about nonconformity, action taken concession obtained, and authority deciding action.

Monitoring, measurement, analysis and evaluation (General) – 9.1.1

- Retain documented information as evidence of the result.

Internal Audit – 9.2–9.2.2

- Retain documented information as evidence of the implementation of the audit program and the audit results.

Management Review – 9.3.3

- Retain documented information as evidence of the management review.

Nonconformity and Corrective Action – 10.2–10.2.2

- Retain documented information as evidence of nature of nonconformity action taken and result of action taken.

5 Checklists

OVERVIEW

The final step of an audit preparation is to generate a checklist. A quality audit checklist is a quality questionnaire record that tracks the planned questions and responses during a quality audit. The quality audit is a valuable tool for continuous improvement. Audits ensure your quality assurance system is sound. Audits are also necessary for ALL ISO registration.

An ISO audit checklist is a key element in planning for and carrying out a process audit, which is a requirement of ALL ISO standards. The checklist for any internal quality audit is composed of a set of questions derived from the quality management system standard requirements and any process documentation prepared by the company. The checklist is created in step 2 and used in step 3 of the five main steps in ISO Internal Audit. As an example, the ISO 9001 clause for management review (9.3) inputs requires that management review to include

- Information on results of audits
- Customer feedback
- Process performance and product conformity
- Status of corrective and preventive actions
- Follow-up actions from previous management reviews
- Changes that could affect the quality management system
- Recommendations for improvement.

If the company process requires that management reviews produce minutes of meeting as a record, then the internal audit checklist could request that the auditor review the minutes of meetings and question that each piece of input information was presented to the management review meeting for assessment.

As this would only be one question on a checklist for reviewing the management review (9.3.1.1–9.3.3.1) process, any ISO audit checklist would contain the many questions required to assess the process. However, it must be noted that sometimes as part of the audit (internal or external), it is possible not to use the prepared checklist because of "new trails" that may lead you to "new" and unexpected paths in your investigation.

In any case, the goal of using checklists for any of the ISO and/or industry standards as well as customer-specific requirements is to review the process and to confirm and validate that the process records provide evidence that the process meets its requirements. The auditor raises an issue for corrective and preventive action ONLY when the process does not meet requirements. It is never used as a punitive action or revenge.

No one will argue that audits may be used as a preventive methodology for improvement. In fact, the way the questions are phrased and delivered to the auditee will define *the* excellent audit. Therefore, audit questions are a critical element of any successful audit including a process, product, or layered process audit (LPA) programs. These questions may be developed with a formal checklist or depend on the experience of the auditor to develop trails and uncover trouble spots along the way. Most organizations develop good or conforming process audit questions that verify the proper and applicable conformance and/or compliance to a particular standard or customer requirement. So, the question for what is an excellent audit is the integrity of the question being asked and who and how the question is being delivered. In other words, in its most fundamental definition of an *excellent audit* is the notion of *value*. However, in no uncertain terms this value must be associated with the improvement of the process. Otherwise, it is an exercise of predetermined inspection practices rather than systematic evaluation. When the former is focused, the auditor is in name only because the emphasis is on the product and not the process. On the other hand, if the focus is on a systematic evaluation, then the auditor is probing with a specific goal to eliminate or reduce waste – make the process efficient. If the outcome of the audit is not improvement, then it is not worth the time, effort, and cost associated with it.

I specifically remember a seminar that I attended long time ago with Peter Drucker in the early 1980s discussing the issues of management and leadership. Some of the items that he mentioned were and still hold true today. One of the most memorable – for me – was the saying: *Management is doing things right; leadership is doing the right things.* How is this relate to auditing? Well, management is in charge of the organization and its resources, which means *efficient utilization*. However, leadership is setting up the vision and values of the organization in order to fulfill the *customer requirements.*

It is of paramount importance to recognize that efficiency and effectiveness are two completely different things. The first is relating to internal requirements – allocation of resources, and the second is relating to customer's needs, wants, and expectations. In both areas, the function of the audit is to find *good* and *bad* practices for improvement. The good ones need to be praised and followed, and the bad ones have to be fixed or improved, given the current technology and budget. Both of them are depended upon the responsibility of management and its leadership to be carried out.

Therefore, before we begin the discussion of specific questions, let us focus on the critical characteristics of defining excellent questions. Based on our experience, we have identified five:

1. To accommodate all standards and requirements, the questions must focus on high risk areas. These may be processes or departments or both. The more risks one identifies, the better the chance for eliminating waste and contributing to improvement.
2. The use of cross-functional and multi-discipline teams in any audit preparation is of paramount importance. The primary focus should be to include individuals that work in the process or department being audited. The closer to the Gemba (actual place – the source) individuals are, the more accurate

the findings will be and received. The reason for their participation is that their particular knowledge of the specific process and/or department is essential to improvement. Typical mistakes that may result because of lack of the appropriate and applicable information or importance may be identified as

- Insufficient or nonexistent definition of what the process is supposed to go. What is its goal?
- Insufficient or nonexistent documentation for the process. Lack of standardization.
- Missing safety requirements to avoid and/or mitigate risks.
- Missing and/or confusing instructions for startup or first piece production.
- Missing and/or lack of appropriate and applicable training for the employees.

So, how can we minimize or even eliminate some of these? We recommend the following program: Use a cross-functional team of *experts* to develop audit questions. We recommend including team members who

- Develop the process in real terms (the *now* process as opposed to *should be* or *could be*).
- Manage process area.
- Provide appropriate and applicable training for those who will do the task(s) in the process.
- Maintain the equipment in the processes. Provide an accurate OEE (Overall Equipment Effectiveness) and preventive maintenance for all equipment. Make sure their capability (P_{pk}) is higher than 1.33. Warning: if the OEE is greater than 100%, something is wrong – usually *the tack (estimated) or takt (actual) time*.
- Before the actual audit, make sure you have a pilot audit to double-check the process and the appropriate questions have been planned.

3. Whatever you do as an auditor, DO NOT USE the language of the standard or the specific requirement. DO NOT USE jargon. AVOID AT ALL COSTS THE USE of acronyms. Rather use the language that employees understand and is meaningful to them. You are not trying to impress anyone. Your function as an auditor is to identify inefficiencies, risks, and excellent processes.

4. Avoid questions that start with WHO. That implies that someone is guilty. That is a wrong approach because the person who is identified is not going to be helpful rather that person will be defensive and reflect something else. The best way to approach the questions is to avoid a YES or No responses. Rather use probing open-ended questions without any insinuations. Therefore, structure questions in a WHAT, HOW, WHY, and REACTION PLAN format:

- WHAT do you want to control, monitor, or evaluate?
- HOW do you audit the WHAT? What makes you sure that the how is accurate? Is it substantiated?

- WHY are you auditing the WHAT? Is it a risk issue? Is it a production issue? Or is it capability issue? Depending on what the issues are, appropriate documentation must be available.
- REACTION PLAN for a nonconforming condition. Continual improvement is based on two issues. The first is *corrective action* which is after the fact. The problem has already occurred and now someone is looking for fixing it. The second – which more often than not, is neglected – is the action that one takes to *prevent* this particular problem never to occur again.

The intent of the questions is in no uncertain terms to *establish* questions that monitor ONLY those operational steps (value-added) which impact or control the quality or the risks of the process. This, of course, takes into account both internal quality and external quality, risk, and manufacturing concerns for each process. When all these items are considered, *presumably* we have a complete checklist and are ready to start the audit.

Noe (2017) reminds us that once you have developed a complete audit checklist, **STOP** and assure the audit

- Is defined in a manner that can be understood at any level of the organization.
- Is focused on process performance and **NOT** product checks, operator performance or QMS system conformance.
- Conforms to all QMS and customer-specific process audit requirements.
- Will assist in continual improvement of your operational performance.

All checklists are generally of the form shown in Table 5.1.

GENERATING A CHECKLIST

In this chapter, we present the reader with thoughts on how to generate a good checklist. Our intent is to give a broad view of the ISO 9001, so that all industries may use this approach. We are also providing the reader with four distinct items. The first, a generic list, forms the basis for generating the questions for the checklist; the second identifies some specific considerations for the environmental standards, the third is a general list for the automotive industry, and the fourth is a checklist for the OH&S standard. None of the checklists are exhaustive; rather, they all provide the impetus for generating questions in specific areas in

---

TABLE 5.1

A Typical Checklist Form for Audits by the Auditor

A Typical Checklist

Standard clause	Reference in your system	Auditor	Verification or validation	Area of concern (possible of nonconformance)	Disposition (yes or no, approved or not approved)

preparation for an ISO audit. Here, we must mention that the IATF-16949 auditor does not have much flexibility to form questions during the audit because the QSA document serves as the checklist.

A Thematic Approach to a Generic Checklist

The team may consider the following items to generate a comprehensive checklist for the organization about to be audited for ISO certification. This is an example of issues that should be addressed and not an exhaustive list.

General

1. Management participation in the program.
2. Authority and responsibility of those in charge of the quality assurance (QA) program are clearly established and documented to show lines of communication for individuals and groups involved.
3. QA organization responsibility includes review of written procedures and surveillance of activities affecting quality.
4. Person or organization responsible for defining and measuring overall effectiveness of the QA program is designated, is independent from pressures of production or testing, has direct access to responsible management, and reports regularly on the effectiveness of the QA program to management.
5. QA manual includes statement of policy and authority of the QA manager, indicates management support, and calls for periodic review of the QA program by a level of management higher than the QA manager.
6. QA manual covers all applicable requirements of referenced standards (if it exists).
7. QA manual revision control system is defined (if it exists).
8. Provision is included for review and approval of the QA manual to assure that it is current and that there is evidence of this approval (if it exists).
9. Identification of the activities, services, and items to which the QA manual applies.
10. QA procedures are available to personnel as required.
11. Provision is included for submission of a controlled copy of the proposed revisions for acceptance.

Design Control

12. Adequate review and comment by applicable groups to assure correct translation of design or test specification requirements into specifications, drawings, procedures, and instructions.
13. Design or test reviews and checking performed by individuals or groups other than those who performed the original design or test.
14. Management of responsible design or testing organization reviews report.
15. System for handling design or test changes provides for acceptance by the organization that performed the design of test.
16. System for communicating design or test changes that affect form, fit, or function to the operators.

Procurement Control

17. Procedures for qualifying suppliers/vendors by survey or appropriate third-party registration:
    a. System to periodically evaluate and revise accepted supplier/vendor list for materials, items, or services being purchased (or alternate approach).
    b. System for removal of suppliers/vendors from list.
    c. Suppliers/vendors identified by name, address, and scope of work or product.
    d. Survey and audit documentation.
    e. Source surveillance/inspection including report system.
    f. Audit frequency.
    g. Corrective action system.
18. Measures to control issuance and approval of purchase documents, including review by QA to assure inclusion of appropriate quality provisions.
19. Receiving inspection.
    a. System providing for use of applicable specifications, purchased drawings, *etc.*
    b. Procedures to assure all required characteristics are reported with recorded results.
    c. Status indicator system and procedures for identification, segregation, and disposition.
    d. Nonconformity control system.
    e. Documentation reports and records to be forwarded to required recipients.
20. Criteria for determining that required activities have been satisfactorily accomplished are included; objective evidence of each accomplishment must be available.
21. Measures established for identification and control of materials and items to assure use of accepted materials and items only.
22. Identification maintained on items or records traceable to item marks or travelers not detrimental to item.
23. Provide for document number and revision to which tests are made, sign-off, and date.
24. Control features of referenced procedures and instructions for quality systems are included in the QA manual.
25. In-process and final examinations and tests established.
26. Review, approval, and revision system documented.

Note: Whenever possible, the organization should deal with suppliers as opposed to vendors. In a supplier relationship, everyone operates in a win-win mode, but in a vendor relationship, the win-lose attitude prevails. A customer works closely with a supplier for the benefit of both organizations, but in a vendor relationship, both parties are interested in getting their own "best deal."

Document Control

27. Measures established to assure revisions and approval are performed prior to release.
28. Issuance of documents that prescribe activities affecting quality is performed by designated personnel using procedures that include accountability, as necessary.
29. Control of revisions and replacements to assure use of current revision and to prevent misuse of obsolete documents.

Control of Processes

30. Process control.
    a. Preparation and control of traveler, process sheet, checklist, etc. in use.
    b. Provision on traveler, *etc.* for establishment of hold point.
    c. Provision on traveler, *etc.*, for signature, initial, or stamp and for date by QA customer representatives for activities is witnessed.
    d. Critical processes including welding, heat treating, NDT (non-destructive testing), forming, and bending accomplished according to referenced specifications and standards using qualified personnel; control for certification of equipment where required.
    e. Provision for accomplishing activities under suitably controlled conditions: appropriate equipment, correct environmental conditions, and prerequisites satisfied in a given activity.

Indoctrination and Training

31. Indoctrination and training program for personnel performing activities affecting quality.
    a. Program description.
    b. Personnel involved.
    c. Documentation records.

Inspection

32. Regular first-line examinations will be verified with quality assurance checks by persons other than those who performed the work or test being inspected.
33. Checklists prepared include document number and revision to which examination is to be performed, provides for recording of results, sign-offs, and dates.

Test Control

34. Written test procedures incorporating or referencing requirements and acceptance criteria from design or testing documents.
35. Test procedures provide for meeting prerequisites, adequate instrumentation, and necessary monitoring. Prerequisites include
    a. Instrumentation calibrated
    b. Personnel trained

    c. Adequate test equipment and items ready to be tested

    d. Data accumulation techniques.

36. Test results documented and evaluated including issue and date of test specification to which test is performed.

Control of Measuring and Test Equipment

37. Measures established and documented to assure use of proper tools, gauges, instruments, and other equipment affecting accuracy quality.

38. Documented procedures established to assure that measuring devices are calibrated and adjusted at specified periods to maintain accuracy.

39. Calibration traceable to national standards or designated if national standards do not exist.

40. Control methods for identifying test equipment and calibration status.

41. Nonconformities in examination and testing equipment.

    a. Equipment identified and findings documented.

    b. Required corrective action

    c. Materials and items checked since previous valid calibration, identified, and considered unacceptable until acceptance established.

    d. Corrective action documented.

Inspection, Test, and Operating Status

42. Measures established to control items per established instructions, procedures, or drawings.

43. Procedures reviewed and approved.

44. Examination and test status control system.

    a. Provides for identification of conforming and nonconforming items.

    b. Provides for control of status indicators.

    c. Indicates authority for application and removal of status indicators.

Nonconformance Control

45. Nonconforming materials or items.

    a. Approved procedures developed for identification, documentation, segregation, and disposition.

    b. Review procedure established for nonconforming items to determine acceptance, rejection, repair, or rework.

    c. Procedures established for accomplishing designated disposition of nonconforming items.

    d. Responsibilities and authority designated for control of nonconformities.

    e. Procedure provides for review by responsible individual in group.

    f. Provision for documentation of disposition and procedures followed in correcting nonconformity or otherwise disposing of items.

    g. Quality records indicate clearly the status of nonconformities.

46. Corrective action.

    a. System provides for prompt identification of nonconformities and determination of recurring nonconformities.

    b. Management awareness and involvement.

    c. Problem, cause of problem, and corrective action implementation documented.

    d. Corrective action program applicable for manufacturer and subcontractors.

Quality Assurance Records

47. Written procedures established to ensure a quality record system.
48. Required records determined early and list developed to serve as index to the document file.
    a. Existence of control documentation.
    b. Existence of master file.
    c. Retrievability of files.
49. Quality program and procedure identifies responsibility for and location of records.
50. System provides suitable protection of records from deterioration and damage.
51. System provides for retention of required records – permanent and nonpermanent – for times specified.
52. Records are reviewed for completeness and review is documented.

Quality Assurance Audit

53. Quality procedures require performance of scheduled audits to assure maintenance of the quality system.
54. Audits are performed to written procedures or checklist applicable for area audited.
55. Audit checklists address areas related to item(s) and system(s) of subcontractors.
56. Management reviews (9.3) are performed and documented to verify that the QA program is functioning.
57. Audit findings and management reviews are documented and reviewed and specific responsibility for corrective action(s) and re-audit is assigned.
58. Reports indicate problems and corrective actions, and show that re-audits verified correction of nonconforming conditions.
59. Personnel performing audits are qualified.
60. Personnel performing audits do not have direct responsibilities in area audited.

Equipment Failure and Malfunction Analysis

61. A system shall be established for handling malfunction and failure reports.
62. A system shall be established for reporting changes to equipment to all purchasers of like equipment.

From the above concerns, it is obvious that a quality audit checklist is a quality questionnaire record that tracks the planned questions and responses during a quality audit. The quality audit is a valuable tool for continuous improvement. Audits ensure your quality assurance system is sound and it is followed. Audits are also necessary for ALL ISO registration.

An ISO audit checklist is a key element in planning for and carrying out a process audit, which is a requirement of ALL ISO standards. The checklist for any internal quality audit is composed of a set of questions derived from the quality management system standard requirements and any process documentation prepared by the company. As an example, the ISO 9001 clause (9.3) for management review inputs requires that management review include:

- Information on results of audits
- Customer feedback
- Process performance and product conformity
- Status of corrective and preventive actions
- Follow-up actions from previous management reviews
- Changes that could affect the quality management system
- Recommendations for improvement.

If the company process requires that management reviews produce minutes of meetings as a record, then the internal audit checklist could request that the auditor review the minutes of meetings and question that each piece of input information was presented to the management review meeting for assessment. As this would only be one question on a checklist for reviewing the management review process, any ISO audit checklist would contain many questions required to assess the process. However, it must be noted that sometimes as part of the audit (internal or external), it is possible not to use the prepared checklist because of "new trails" that may lead you to "new" and unexpected paths in your investigation.

In any case, the goal of using checklists for any of the ISO and/or industry standards as well as customer-specific requirements is to review the process and to confirm and validate that the process records provide evidence that the process meets its requirements. The auditor raises an issue for corrective and preventive action ONLY when the process does not meet requirements. It is never used as a punitive action or revenge.

A Typical Internal Preassessment Survey

The following questions are intended to be used only as guidelines in a given organization. The questions are designed to identify any shortcomings in your system and to allow you to plan accordingly. They are not meant to be used as a formal checklist for any organization, since the official checklist is prepared by the representatives of the registrar. The list is based on Grossman (1995, pp. 34–35).

Does your company have a written quality policy that describes management's commitment to quality and objective for achieving quality in every part of the company's operation?

1. Has your management group endorsed the quality policy and communicated the policy to all employees?
2. Is there an approved organization chart showing who is responsible for all work that affects the quality of the product or service that your company produces?

3. Are the functions and job specifications for personnel who affect the quality of the product or service clearly defined?
4. Are the technical and personnel resources that are needed for the inspection, testing, and monitoring of the production of the product or service made available by management?
5. Are the technical and personnel resources that are needed for the inspection, testing, and monitoring of the product or service during its life cycle made available by management?
6. Are periodic audits of the quality system completed as often as necessary to keep each part of the system in control? (Internal audits – process or layered audits)
7. Are periodic audits of the manufacturing processes completed as often as necessary to keep each process in control?
8. Are periodic audits of the product or service that your company produces completed as often as necessary to ensure that the quality of the product meets customer requirements?
9. Are the results of the audit communicated to management and to those employees who affect quality?
10. Has your company appointed a coordinator to be responsible for monitoring the quality system and calling attention to the deficiencies?
11. Are quality reviews held at appropriate intervals?
12. Are the results of the audits recorded and maintained?
13. Are procedures written for each activity that affects quality? Are they appropriately maintained? Are they easily accessed by the employees?
14. Does your company have a plan for achieving and maintaining quality?
15. Does your company audit and evaluate its progress in achieving the objectives listed in the quality plan?
16. Are customer needs identified and communicated to all employees who affect the quality of the product?
17. Do employees know what they have to do on the job to provide the desired level of quality in the product or service?
18. Are the customer requirements for product and service quality adequately defined in the contract with the customer?
19. Are customer contracts reviewed for accuracy?
20. Are records of the customer reviews maintained?
21. Are incomplete and ambiguous requirements resolved before design or production?
22. Are all applicable and appropriate documents reviewed before they are released for use?
23. Do you have an obsolescent policy? Do you follow it? What is your policy for discarding it?
24. Do your procedures and instructions describe what is actually done on the job, now?
25. Do you have document control? Do you follow it?
26. Do you have a certification program for your suppliers? If not, how do you approve your suppliers?

27. Do you keep performance records from your suppliers? Do you perform regular analysis with the data? Do you communicate the information to your supplier base?
28. Does someone check all incoming supplies and equipment to verify that you have indeed received the correct resources to do the job and that they meet the defined requirements?
29. Do you maintain it list of approved suppliers?
30. Do you audit your suppliers?
31. Do you use systematic methods to identify and plan production processes and (if appropriate) equipment and product installation processes?
32. Do your employees use their own tools? How do you make sure they are calibrated?
33. Do you do calibration?
34. Do you have written set-up and process instructions?
35. Do you have preventive maintenance?
36. Do you have written standards for workmanship and criteria for meeting the standards?
37. Do your employees follow job procedures and instructions?
38. Do your employees follow unwritten procedures or instructions?
39. Do the procedures and instructions describe the way employees do their jobs now?
40. Do you record tooling repairs to ensure process control?
41. Do you have written procedures to ensure that incoming products are not used or handled before an inspection or other form of verification proves that these products meet specified requirements?
42. Are inspection procedures carried out in accordance with written instructions and your company's quality plan?
43. Do you have written procedures to identify incoming material that may have been released before it was inspected because of urgent production purposes?
44. Do you maintain a receiving inspection history or log?
45. Does your company collect and maintain records to prove that you have met customer requirements?
46. Do you have written instructions for inspecting and testing?
47. Are nonconforming products identified and separated so that they are not sent to customers? What is your quarantine policy?
48. Are there written procedures to verify that all final inspections and tests are completed before products are sent to customers?
49. Are there written procedures for calibration and maintenance of inspection, measurement, and test equipment that show calibration frequency?
50. Do you have a system to identify the inspection or test status of products during manufacturing?
51. Is there a documented procedure for identifying and separating rejected material to prevent inadvertent use of nonconforming products?

52. Is there a method of recording the rejected material and the disposition of such material? Are there documents to support that the method is being followed?
53. Is there a method for requesting a deviation from the customer? Is it being followed? Is there documentation to support the practice?
54. When a waiver of change or a deviation has been authorized by the customer, is that information recorded and maintained?
55. Is there an analysis of nonconformities?
56. Are there procedures for ensuring that effective corrective actions are carried out?
57. Are there procedures from preventing damage to products as they are handled?
58. Are in-stock products inspected at periodic intervals?
59. Is there a written procedure for identifying, collecting, indexing, filing, maintaining, and disposing of quality-related records?
60. Are quality records maintained so that the achievement of their required levels of quality can be demonstrated to customers and to your management team?
61. Are quality records stored in an accessible place? Are they retrievable?
62. Are quality records accessible to your customers for their review?
63. Do you have a retention policy? Is it written? Is it being followed?
64. Are quality audits performed as defined in your procedures?
65. Do the appropriate personnel take timely corrective actions? Are their actions recorded?
66. Do training and development plans exist for all employees who have an impact on the organization, product, or service?
67. Are records maintained to show who attended training, when they attended, and their success in learning the skills?
68. Are there written procedures and instructions for follow-up service? Does appropriate maintenance exist for these procedures? Do they meet the requirements?
69. Is there a method of establishing the need for statistical techniques? How do you maintain control in your processes?

The reader of this checklist should notice that the accepted answer for all questions is a "YES" response. This is very shallow respond and in fact in most cases a questionable response. Therefore, this list is given here to be used ONLY as the "spring board" for other follow-up questions. The intent is to help the auditor "break the ice" and to help the operator be at ease. In essence, it gives confidence to both the auditee and auditor of a positive experience. It avoids the "got you" moment.

A FORMAL ISO 9001:2015 CHECKLIST

The following checklist can be used for both internal and external audits as well as a gap analysis tools.

CLAUSE 4: CONTEXT OF THE ORGANIZATION

4.1 Understanding the organization and its context

- Has the organization determined the external and internal issues relevant to the purpose and strategic direction of its QMS and that can affect its ability to achieve the intended results?
- Does the organization monitor and review information about these external and internal issues?

4.2 Understanding the needs and expectations of interested parties

- Has the organization determined the interested parties that are relevant to the QMS?
- Has the organization determined the requirements of these interested parties relevant to the QMS?
- Does the organization monitor and review the information about these interested parties and their relevant requirement?

4.3 Determining the scope of the quality management system

- Has the organization established the scope of its QMS?
- Has the organization determined the boundaries and applicability of the QMS?
- While determining the scope, has the organization determined the external and internal issues, requirements of relevant interested parties, and products and services of the organization?
- While determining applicability, does the organization determine if it affects its ability or responsibility to ensure the conformity of its products and services and the enhancement of customer satisfaction?
- Does the scope state the types of products and services covered?
- Does the scope give justification for any requirements that the organization determines and is not applicable to the scope of its QMS?
- Is the organization's scope made available and maintained as a documented information?

4.4 Quality management system and its processes

- Has the organization established, implemented, maintained, and continually improved its QMS?

4.4.1

- Has the organization determined the processes needed for the QMS?
- Has the organization determined the application of these process throughout the organization?

- Has the organization determined the sequence and the interaction of these process?
- Has the organization determined and applied the criteria and methods needed to ensure the effective operation and control of these processes?
- Do these methods include the monitoring, measurement, and related performance indicator?
- Has the organization determined the resources needed for these organization?
- Has the organization ensured the availability of the resources needed for these processes?
- Has the organization assigned the responsibilities and authorities for these processes?
- Has the organization addressed the risk and opportunities associated with these processes?
- Has the organization evaluated these processes and implemented any changes needed to ensure that these processes achieve its intended results?
- Has the organization made improvement in its processes and its QMS?

4.2.2

- Has the organization maintained documented information to support the operation of its processes?
- Do the organization retain documented information as an evidence that the processes have been carried out as planned?

CLAUSE 5: LEADERSHIP

5.1 Leadership and commitment
  5.1.1 General

- Does the top management demonstrate leadership and commitment by taking accountability for effectiveness of its QMS?
- Has the top management ensured that the quality policy and quality objective are established?
- Are quality policy and quality objective compatible with the context and strategic direction of the organization?
- Has the organization integrated the requirements of QMS with the business processes?
- Is the organization promoting the use of process approach and risk-based thinking throughout the organization?
- Is the top management ensuring that the resources needed for the QMS are available?
- Is the importance of effectiveness of QMS and meeting QMS requirements communicated?
- Does the top management ensure that the QMS is achieving its intended results?

- Does top management engage, direct, and support the persons required to contribute to the effectiveness of the QMS requirements?
- Is top management promoting improvements through continuous improvement in the process(es)?
- Is top management supporting other relevant management roles to demonstrate their leadership as it applies to their area of responsibilities?

5.1.2 Customer focus

- Does the top management demonstrate leadership and commitment by ensuring that customer and applicable statutory and regulatory requirements are determined, understood, and consistently meeting the requirements?
- Are the risks and opportunities that can affect conformity of products and services and the ability to enhance customer satisfaction determined and addressed?
- Is the focus of enhancing customer satisfaction maintained?

5.2 Policy
5.2.1 Establishing the quality policy

- Has the top management established, implemented, and maintained a quality policy?
- Is quality policy appropriate to the purpose and context of the organization and does it support its strategic direction?
- Does quality policy provide the framework for setting quality objectives?
- Does quality policy include the commitment to satisfy applicable requirements and to continually improvement of the QMS?

5.2.2 Communicating the quality policy

- Is quality policy maintained as a documented information?
- Is quality policy communicated, understood, and applied within the organization?
- Is quality policy appropriate and made available to the relevant interested parties?

5.3 Organizational roles, responsibilities, and authorities

- Has the top management ensured that the responsibilities and authorities for relevant roles are assigned, communicated, and understood within the organization?
- While assigning the responsibilities and authorities, does the top management ensure that its QMS meets the requirement of ISO 9001:2015?
- While assigning the responsibilities and authorities, does the top management ensure that the processes are meeting its intended results?

- While assigning the responsibilities and authorities, does the top management ensure that there is promotion of customer focus throughout the organization?
- While assigning the responsibilities and authorities, does the top management ensure that performance of its QMS and opportunities for improvement are reported to them?
- While assigning the responsibilities and authorities, does the top management ensure that integrity of QMS is maintained when changes to the QMS are planned and maintained?

CLAUSE 6: PLANNING

6.1 Actions to address risks and opportunities

- While planning for QMS, does the organization considers the issues referred in clause 4.1 and requirement referred in clause 4.2?

6.1.1

- Has the organization determined the risks and opportunities that has to be addressed so that QMS can achieve its intended results, enhance desirable effects, prevent, or reduce undesired effects and achieve improvement?

6.1.2

- Has the organization planned actions to address these risks and opportunities?
- Have these actions implemented and integrated into its QMS processes?
- Has the organization evaluated the effectiveness of these actions?
- Is the action proportionate to the potential impact on the conformity of product and services?

6.2 Quality objectives and planning to achieve them
    6.2.1

- Has the organization established quality objectives at relevant functions, levels, and process needed for the QMS?
- Are the quality objectives consistent with the quality policy?
- Does the organization have quality objectives which are relevant to the conformity of product and services and enhancement of customer satisfaction?
- Are quality objectives measurable and do they take account of applicable requirements?
- Are quality objectives monitored, communicated, and updated as required?
- Does the organization maintain documented information on the quality objectives?

6.2.2

- For achieving quality objectives does the organization determine what will be done, what resources are required, who will be responsible, when will it be completed, and how are the result to be evaluated?

6.3 Planning for change

- While determining changes for the QMS, are changes carried out in planned manner?
- While planning for change, does the organization consider the purpose of the change and their potential consequence; the integrity of the QMS; the availability of resources; and allocation and reallocation of responsibilities and authorities?

CLAUSE 7: SUPPORT

7.1 Resources
   7.1.1 General

- Has the organization determined and provided the resources needed for establishing, implementing, maintaining, and continually improving the QMS?
- Has the organization considered the capabilities and constraints of existing internal resources?
- Has the organization considered what needs to be obtained from external providers?

7.1.2 People

- Has the organization determined and provided the persons required for effective maintenance of QMS and for operation and control of its processes?

7.1.3 Infrastructure

- Has the organization determined and maintained the infrastructure needed for operation of its processes and to achieve conformity of product and services?

7.1.4 Environment for the operation of processes

- Has the organization determined, provided, and maintained the environment necessary for the operation of its processes and to achieve conformity of products and services?

7.1.5 Monitoring and measuring resources
   7.1.5.1 General

- Has the organization determined and provided the necessary resources needed when monitoring and measuring is used to verify conformity to product and service requirement?
- Are resources suitable for the type of monitoring and measurement activities undertaken?
- Are resources maintained to ensure their continuing fitness?
- Does the organization retain appropriate documented information as evidence of fitness for the purpose of the monitoring and measurement resources?

7.1.5.2 Measurement traceability

- Is there a requirement for measurement traceability?
- Where measurement traceability is a requirement, are measurement equipment calibrated or verified at specified interval or prior to use?
- Is the calibration completed against measurements or standards traceable to national or international standards?
- Where no such standard is existing, are documented information retained for the basis used for calibration or verification?
- Are measuring equipment identified in order to determine their status?
- Are measuring equipment safeguarded from adjustments, damages, or deteriorations that would invalidate the calibration and subsequent measurement results?
- Does the organization determine and take appropriate action if the validity of pervious measurement results has been adversely affected when measuring equipment is found to be unfit for its intended purpose?

7.1.6 Organizational knowledge

- Does the organization determine the knowledge necessary for the operation of its processes and to achieve conformity of product and services?
- Does the organization maintain this knowledge and make it available to the extent necessary?
- While addressing changing needs and trends, does the organization consider its current knowledge and determine how to acquire or access any necessary additional knowledge and required updates?

7.2 Competence

- Does the organization determine the necessary competence of its employees whose work affects the performance and effectiveness of the QMS?
- Does the organization ensure that its employees are competent on the basis of appropriate education, training, or experience?

- Does the organization take applicable actions to acquire the necessary competence and evaluate the effectiveness of action taken?
- Does the organization retain the appropriate documented information as evidence of competence?

7.3 Awareness

- Does the organization ensure that the persons doing work under the organization's control are aware of its quality policy, relevant quality objectives, their contribution to the effectiveness of QMS including the benefits of improved performance, and the implications of not meeting QMS requirements?

7.4 Communication

- Does the organization determine the internal and external communication relevant to the QMS including on what it will communicate, when to communicate, with whom to communicate, how to communicate, and who communicates?

7.5 Documented Information
7.5.1 General

- Does the organization's QMS include documents required by ISO 9001:2015 and documents determined by the organization necessary for the effectiveness of the QMS? (See Note 1.)

7.5.2 Creating and updating

- While creating and updating documented information, does the organization ensure it is appropriate in terms of identification and descriptions?
- While creating and updating documented information, does the organization ensure that it is in proper format and in correct media?
- While creating and updating documented information, does the organization ensure that there is appropriate review and approval for suitability and adequacy?

7.5.3 Control of documented information
7.5.3.1

- Does the organization control its documented information to ensure that it is available and suitable for use, whenever it is needed?
- Is the documented information adequately protected?

7.5.3.2

- Is the distribution, access, retrieval, and use of documented information adequately controlled?

- Is the documented information properly stored and adequately preserved and is it legible?
- Is there control of changes (e.g., version control)?
- Are adequate controls in place for retention and disposition?
- Is external origin documented information necessary for planning and operation of QMS appropriately identified and controlled?
- Are records protected for unintended alterations?

CLAUSE 8: OPERATIONS

- Does the organization plan, implement, and control the processes needed to meet the requirement for the provision of product and services and to implement the action determined in clause 6?

8.1 Operation planning and control

- Does the organization determine the requirements for the products and services?
- Has the organization established criteria for the processes and acceptance of products and services?
- Does the organization determine the resources needed to achieve conformity to the product and service requirements?
- Does the organization implement controls of the processes in accordance with the criteria?
- Does the organization determine, maintain, and retain necessary documented information to have confidence that the processes have been carried out as planned and to demonstrate the conformity of products and services?
- Does the organization control its planned changes and review the consequences of unintended changes?
- Does the organization take action to mitigate any adverse effects of its unintended changes?

8.2 Requirements for products and services
8.2.1 Customer communication

- Does the organization communicate with customers to provide information relating to products and services, handling enquiries, contracts, or orders (including any changes)?
- Does the organization obtain customer feedback relating to products and services including customer complaint?
- Does the organization communicate with the customers relating handling or controlling customer property?
- Has the organization established requirements for contingency action, where required?

8.2.2 Determining the requirements for products and services

- Has the organization determined the requirements for product and services to be offered the customer?
- Are the requirements defined and do they include applicable statutory and regulatory requirements and those considered necessary by the organization?
- Can the organization meet the claims for the product and services it offers?

8.2.3 Review of the requirements for products and services

- Has the organization ensured that it has the ability to meet the requirements for products and services?
- Has the organization conducted a review before committing to supply product and services?
- Has the organization reviewed the requirements specified by customer, including the requirements for delivery and post-delivery activities?
- Has the organization reviewed the requirements not stated by the customers but necessary for the specified or intended use when know?
- Has the organization reviewed the statutory and regulatory requirements applicable to the product and services and requirements specified by the organization?
- Has the organization reviewed and resolved contract or order requirements differing for those previously defined?
- When the customer does not provide a documented statement of their requirement, does the organization confirm to the customer's requirements before acceptance?
- Does the organization retain documented information on the results of the review and on any new requirements for the products and services?

8.2.4 Changes to requirements for products and services
- Does the organization ensure that the relevant documented information is amended and the relevant persons are made aware of the changed requirements, when the requirements for the products and services are changed?

8.3 Design and development of products and services
  8.3.1 General
- Has the organization established, implemented, and maintained a design and development (D&D) process that is appropriate to the subsequent provision of product and services?

8.3.2 Design and development planning

- In determining the stages and controls for D&D, has the organization taken into consideration the nature, duration, and complexity of D&D activities?
- In determining the stages and controls for D&D, has the organization taken into consideration the required process stages including D&D reviews?

- In determining the stages and controls for D&D, has the organization taken into consideration the D&D verification and validation activities?
- In determining the stages and controls for D&D, has the organization taken into consideration the responsibilities and authorities involved in the D&D process?
- In determining the stages and controls for D&D, has the organization taken into consideration the external and internal resources needed?
- In determining the stages and controls for D&D, has the organization taken into consideration the need to control interfaces between persons involved in D&D?
- In determining the stages and controls for D&D, has the organization taken into consideration the need for involvement of customer and user?
- In determining the stages and controls for D&D, has the organization taken into consideration the requirements of subsequent provision of product and services?
- In determining the stages and controls for D&D, has the organization taken into consideration the level of the control expected for the D&D by customers and other relevant interested parties?
- In determining the stages and controls for D&D, has the organization taken into consideration the documented information needed to demonstrate that D&D requirement have been met?

8.3.3 Design and Development inputs

- Has the organization determined the essential requirements for the specific types of products and services to be designed and developed?
- Does the organization consider the following functional and performance requirements; statutory and regulatory requirements; standards or code of practices that the organization has committed to implement; information derived from previous D&D activities; and potential consequences of failure due to the nature of the product and services?
- Does the organization ensure that the inputs are adequate for D&D purpose, complete, and unambiguous?
- Does the organization resolve the conflicting D&D inputs?
- Is the documented information for D&D inputs retained?

8.3.4 Design and development controls

- Has the organization applied the necessary controls to D&D processes to ensure that the result to be achieved are defined?
- Has the organization conducted review to evaluate the ability of the results of D&D to meet the requirements?
- Has the organization conducted the verification to ensure that D&D meets input requirements? Is geometric dimensioning and tolerancing (GD&T) used? Are the appropriate employees trained for it?

- Has the organization conducted the appropriate and applicable validation to ensure that the resulting product and service meets the requirements of the specified application or intended use?
- Has the organization taken necessary action on the problems determined during reviews, verification, or validation activities?
- Has the organization retained documented information on the abovementioned activities?

8.3.5 Design and development outputs

- Does the organization ensure that D&D outputs meets the input requirements?
- Does the organization ensure that D&D outputs are adequate for the subsequent processes for provision of product and services?
- Does the organization ensure that D&D outputs include (or has reference) monitoring and measuring requirements and acceptance criteria?
- Does the organization ensure that D&D outputs specify the characteristics of the products and services that are essential for their intended use?

8.3.6 Design and development changes

- Has the organization identified, reviewed, and controlled changes made during, or subsequent to the D&D of the product and services to ensure that there is no averse to the impact on conformity to requirement?
- Has the organization retained the documented information on D&D changes, the result of reviews, authorization of the changes, and the action taken to prevent adverse impact?

8.4 Control of externally provided processes, products, and services
8.4.1 General

- Does the organization ensure that the externally provided processes, products, and services conform to the requirements?
- Does the organization determine the controls needed when the product and services from the external providers are incorporated into their own product and services?
- Does the organization determine the controls needed when the product and services from the external providers are provided directly to the customer by external providers?
- Does the organization determine the controls needed when process or part of process is provided by the external providers?
- Has the organization determined and applied the criteria for selection, evaluation, monitoring of performance, and re-evaluation of external providers?
- Has the organization retained the documented information of these activities and any action arising out or evaluation/re-evaluation?

8.4.2 Type and extent of control

- Does the organization ensure that the externally provided processes, products, and services do not adversely affect its ability to consistently deliver conforming products and services to the customers?
- Does the organization ensure that the externally provided process remains within the control of its QMS?
- Has the organization defined the controls to be applied to external provider and its resulting outputs?
- Has the organization taken into consideration the potential impact of the organization's ability to consistently meet customer and applicable statutory and regulatory requirement?
- Has the organization taken into consideration the effectiveness of the controls applied by the external providers?
- Has the organization determined the verification or other activities, necessary to ensure that the externally provided processes, products, and services meet the requirements?

8.4.2 Information for external providers

- Does the organization ensure adequacy of requirements prior to their communication to the external provider?
- Does the organization communicate to the external providers its requirements for the processes, products, and services required?
- Does the organization communicate to the external providers its requirements for the approval of the products and services; methods, processes, and equipment; the release of products and services?
- Does the organization communicate to the external providers its requirements for competence including any qualification of persons?
- Does the organization communicate to the external providers its requirements for external provider's interactions with the organization?
- Does the organization communicate to the external providers its requirements for control and monitoring of the external providers' performance to be applied by the organization?
- Does the organization communicate to the external providers its requirements for verification or validation activities that the organization or its customer intends to perform at the external providers' premises?

8.5 Production and Service provision
8.5.1 Control of production and service provision

- Has the organization implemented production and service provision under controlled conditions?
- Is there any documented information available that defines the characteristics of the product, services, or activities to be performed and the results to be achieved?

- Are any suitable monitoring and measuring resources available? Are they being used?
- Are monitoring and measuring activities being performed at appropriate stages?
- Are competent persons (including qualification) being appointed?
- Is the infrastructure and environment being used suitable for operation of processes?
- Has the organization implemented any actions to prevent human error?
- Has the organization implemented any release, delivery, and post-delivery activities?
- Where resulting output cannot be verified by subsequent monitoring or measurement, has the organization conducted validation and periodic revalidation of the process for production and service provision?

8.5.2 Identification and traceability

- Has the organization used any suitable means to identify output when it is necessary to ensure the conformity of products and services?
- Has the status of outputs with respect to monitoring and measuring requirements throughout the production and service provision being identified by the organization?
- Has the organization controlled the unique identification of the outputs when traceability is a requirement?
- Has the organization retained the documented information necessary to enable traceability, when traceability is a requirement?

8.5.3 Property belonging to customers or external providers

- When property belonging to customers or external providers is under the organization's control or being used by the organization, does the organization exercise adequate care?
- Does the organization identify, verify, protect, and safeguard customers' or external providers' property?
- When the property or the customer or external provider is lost, damaged or otherwise found to be unsuitable for use, does the organization report this to the customer or external provider? Does the organization retain documented information on what has occurred?

8.5.4 Preservation
- Does the organization preserve the outputs during production and service provision, to the extent necessary to ensure conformity to requirements?

8.5.5 Post-delivery activities

- Does the organization meet requirements for post-delivery activities associated with the product and services?

- In determining the extent of post-delivery activities, does the organization consider the statutory and regulatory requirements; the potential undesired consequences associated with its product and services; customer requirement and feedback; the nature, use, and intended lifetime of its product and services?

8.5.6 Control of change

- Does the organization conduct review and control changes for production or service provision to ensure continuing conformity with requirements?
- Does the organization retain documented information describing the results of the review of changes, the person(s) authorizing the change, and any necessary actions arising from the review?

8.6 Release of products and services

- Has the organization implemented planned arrangements, at appropriate stages, to verify that the product and service requirements have been met?
- Does the organization ensure that the release of product and service proceeds only after the planned arrangement is satisfactorily completed or approved by relevant authority and as applicable by the customer?
- Does the organization retain the documented information on the release of products and services and include information relating to the evidence of conformity with the acceptance criteria and traceability of the person authorizing the release?

8.7 Control of nonconforming outputs
8.7.1

- Does the organization ensure that the outputs which do not conform to their requirements are identified and controlled to prevent their unintended use or delivery?
- Is the action appropriate to the nature of the nonconformity and its effect on the conformity of product and services?
- Do the organization also consider nonconforming product and services detected after delivery of products, during and after provision of services?
- When nonconforming products and services are detected, does the organization take correction action and/or segregation, containment, return, or suspension of provision of products and services and/or informing the customer and/or obtaining authorization for acceptance under concession?
- Does the organization retain documented information that describes the nonconformity, describes the actions taken, describes any concession obtained, and identifies the authority deciding the action in respect of the nonconformity?

CLAUSE 9: PERFORMANCE EVALUATION

9.1 Monitoring, measurement, analysis, and evaluation
   9.1.1 General

- Did the organization plan how to monitor, measure, analyze, and evaluate its QMS?
- Did the organization plan how to monitor QMS performance and effectiveness?
- Did the organization figure out what needs to be monitored and select methods? How? Why?
- Did the organization determine its QMS monitoring requirements?
- Does the organization select monitoring methods that can produce valid results?
- Did the organization establish when monitoring should be done and who should do it?
- Did the organization plan how to measure QMS performance and effectiveness?
- Did the organization figure out what needs to be measured and did the organization select methods? Did the organization determine its QMS measurement requirements?
- Does the organization select measurement methods that can produce valid results? Did the organization establish when measuring should be done and who should do it?
- Did the organization plan how to analyze QMS performance and effectiveness?
- Did the organization select analytical methods that are capable of producing valid results?
- Did the organization decide when monitoring and measurement results are analyzed?
- Did the organization plan how to evaluate QMS performance and effectiveness?
- Did the organization select evaluation methods that are capable of producing valid results?
- Did the organization decide when monitoring and measurement results are evaluated?
- Does the organization monitor, measure, analyze, and evaluate the organization organization's QMS?
- Does the organization monitor the performance and effectiveness of the organization's QMS?
- Does the organization record monitoring results and retain and control these records?
- Does the organization measure the performance and effectiveness of the organization's QMS?
- Does the organization record measurement results and retain and control these records?

- Does the organization analyze the performance and the effectiveness of its QMS?
- Does the organization record analytical results and do the organization retain and control these records?
- Does the organization evaluate the performance and effectiveness of its QMS?
- Does the organization record the evaluation results, and retain and control these records?

9.1.2 Customer satisfaction

- Does the organization establish methods that the organization can use to monitor customer perceptions?
- Does the organization figure out how the organization is going to obtain information about how customers feel about how well it is meeting their needs and expectations?
- Does the organization figure out how the organization is going to review information about how customers feel about how well it is meeting their needs and expectations?
- Does the organization monitor how well customer needs and expectations are being fulfilled?
- Does the organization monitor how the organization's customers feel about how well the organization is meeting their needs and expectations (do the organization monitor the organization's customers' perceptions)? What methods do they use for such monitoring?

9.1.3 Analysis and evaluation

- Does the organization analyze its monitoring and measurement results?
- Does the organization analyze and evaluate appropriate data and information?
- Does the organization use its analytical results to evaluate performance?
- Does the organization evaluate the performance of its QMS?
- Does the organization determine if it needs to improve its performance?
- Does the organization evaluate the performance of its external providers?
- Does the organization use its analytical results to evaluate effectiveness?
- Does the organization evaluate the effectiveness of its QMS?
- Does the organization determine if it needs to improve its effectiveness?
- Does the organization evaluate the effectiveness of its planning?
- Does the organization determine if its plans were effectively implemented?
- Does the organization evaluate the effectiveness of its actions?
- Does the organization evaluate the effectiveness of actions taken to address risks?
- Does the organization evaluate the effectiveness of actions taken to address opportunities?
- Does the organization use its analytical results to evaluate conformity?

- Does the organization evaluate the conformity of products and services?
- Does the organization use its analytical results to evaluate satisfaction?
- Does the organization evaluate the degree of customer satisfaction?

9.2 Internal Audit
9.2.1

- Does the organization conduct internal audits at planned intervals?
- Did the organization plan a program that can find out if QMS meets organization's own requirement and ISO 9001:2015 requirements?
- Did the organization plan a program that can find out if QMS is effectively implemented and maintained?

9.2.2

- Did the organization plan, establish, implement, and maintain an audit program?
- Did the audit program include the frequency, methods, responsibilities, planning requirements, and reporting?
- Does the audit program take into consideration the importance of the process concerned, changes affecting the organization, and the results of previous audits?
- Did the organization define the audit criteria and scope of each audit?
- Does the organization ensure that the audit is conducted by the auditors to ensure objectivity and impartiality of the audit process?
- Does the organization ensure that the results of the audits are reported to relevant management?
- Does the organization take appropriate correction and corrective action without undue delays?
- Does the organization retain documented information as evidence of the implementation of the audit program and the audit results?

9.3 Management review
9.3.1 General

- Does the top management review the organization QMS at planned intervals?
- Does the review ensure QMS's continuing suitability, adequacy, effectiveness, and alignment with the strategic direction of the organization?

9.3.2 Management review inputs

- Does the review take into consideration the status of actions from previous management reviews?
- Are the changes in external and internal issues relevant to QMS considered?
- Does the review take into consideration information on the performance and effectiveness of the QMS?

- Does the review take into consideration customer satisfaction and feedback from relevant interested parties?
- Does the review take into consideration the extent to which the quality objectives have been met?
- Does the review take into consideration process performance and conformity of product and services?
- Does the review take into consideration nonconformities and corrective actions?
- Does the review take into consideration monitoring and measuring results?
- Does the review take into consideration audit results?
- Does the review take into consideration the performance of external providers?
- Does the review take into consideration adequacy of resources?
- Does the review take into consideration the effectiveness of actions taken to address risks and opportunities?
- Does the review take into consideration the opportunities for improvement?

9.3.3 Management review outputs

- Do the outputs of the management review include decision and actions related to the opportunities for improvement; any need for changes to the QMS; and resource needed?
- Does the organization retain documented information as evidence of the result of management review?

CLAUSE 10: IMPROVEMENT

10.1 General

- Has the organization determined and selected opportunities for improvement?
- Has the organization implemented any necessary action to meet customer requirement and enhance satisfaction?
- Has the organization taken action for improving products and services to meet requirements as well as to address future needs and expectations?
- Has the organization taken action for correcting, preventing, or reducing undesired effects?
- Has the organization taken action for improving the performance and effectiveness of the QMS?

10.2 Nonconformity and corrective action

- When any nonconformity (including complaints) occurs, does the organization take action to control and correct it and deal with the consequences?
- When any nonconformity (including complaints) occurs, does the organization evaluate the need for action to eliminate the causes of the nonconformity?

- Does the organization review and analyze the nonconformity?
- Does the organization determine the causes of the nonconformity?
- Does the organization determine similar nonconformity exist or could potentially occur?
- Has the organization implemented any action needed?
- Has the organization reviewed the effectiveness of the corrective action taken?
- Has the organization updated risk and opportunities determined during planning if necessary?
- Has the organization made changes to the QMS if necessary?
- Are the corrective actions appropriate to the effects of the nonconformities encountered?

10.2.2 Nonconformity and corrective action

- Does the organization retain documented information on the nature of the nonconformities and any subsequent actions taken, and result of any corrective action?

10.3 Continual improvement

- Does the organization continually improve the suitability, adequacy, and effectiveness of the QMS?
- Does the organization consider the results of analysis and evaluation, and output from management review to determine if there are needs or opportunities to be addressed as part of continual improvement?

Note 1: In Element 7.5.1, it was mentioned "required documents." These are the required records and documents:

ISO 9001:2015 Mandatory Records

- Monitoring and measuring equipment calibration records (clause 7.1.5.1)
- Records of training, skills, experience, and qualifications (clause 7.2)
- Product/service requirements review records (clause 8.2.3.2)
- Record about design and development outputs review (clause 8.3.2)
- Records about design and development inputs (clause 8.3.3)
- Records of design and development controls (clause 8.3.4)
- Records of design and development outputs (clause 8.3.5)
- Design and development changes records (clause 8.3.6)
- Characteristics of product to be produced and service to be provided (clause 8.5.1)
- Records about customer property (clause 8.5.3)
- Production/service provision change control records (clause 8.5.6)
- Record of conformity of product/service with acceptance criteria (clause 8.6)
- Record of nonconforming outputs (clause 8.7.2)
- Monitoring and measurement results (clause 9.1.1)

- Internal audit program (clause 9.2)
- Results of internal audits (clause 9.2)
- Results of the management review (clause 9.3)
- Results of corrective actions (clause 10.1).

ISO 9001:2015 MANDATORY DOCUMENTS

- Scope of the QMS (clause 4.3)
- Quality policy (clause 5.2)
- Quality objectives (clause 6.2)
- Criteria for evaluation and selection of suppliers (clause 8.4.1).

CHECKLIST FOR ISO 14000 – ENVIRONMENTAL STANDARDS

OVERVIEW

ISO 14000 is a standard that focuses in the environment. However, quite a few of the clauses are very similar to the ISO 9001. Because of the similarity, we recommend that the reader review the ISO 9001 checklists that we provided in the previous section. We believe that the questions are quite appropriate, and perhaps with some minor modifications, they can indeed apply to individual organizations trying to meet the requirements of the standard. In addition, we have provided a list of "things to consider" based on the NSF's list provided on their website and used with permission: http://www.nsf.org/newsroom_pdf/isr_changes_iso14001.pdf. Retrieved on November 24, 2019.

The newest version is the ISO 14001:2015 which defines the requirements for an environmental management system (EMS) that an organization can use to enhance its environmental performance. Specifically, the predominant purpose of the ISO 14001:2015 is for use by an organization seeking to manage its environmental responsibilities in a systematic manner that contributes to the environmental pillar of sustainability. This means that the auditor should be able to consider the following when conducting "the" audit:

- Does the organization have a hard or an electronic copy of the standard?
- Have all the appropriate people read it? Did they have the appropriate training?
- Have the appropriate timelines been formed to define the scope, schedule, and budget for the implementation or continuing the ISO 14001 requirements?
- Have appropriate measures been taken for appropriate guidance?
- Is the organization ready for the changes?
- Is the management on board? Do they understand their role?
- Is the life cycle perspective understood?
- How the baselines are measured? How and why were they selected?
- Is the organization ready to identify (a) risks, opportunities, consequences, and a plan for prevention of these risks?

So, any organization that follows the ISO 14001:2015 hopes to achieve the intended outcomes of its environmental management system, which provide value for the environment, the organization itself, and interested parties. Generally speaking, everyone agrees that by following this standard which should be consistent with the organization's environmental policy, the expected results of an environmental management system will include (a) enhancement of environmental performance, (b) fulfillment of compliance obligations, and (c) achievement of environmental objectives.

So, the fact that environmental issues and concerns over the last 10 years have increased exponentially the ISO 14001:2015 standard has become – more or less – the default and applicable standard to any organization, regardless of size, type, and nature. It applies to the environmental aspects of its activities, products, and services that the organization determines and it can either control or influence considering a life cycle perspective. It is very interesting to note that the ISO 14001:2015 does not state specific environmental performance criteria.

Again, just like in the ISO and IATF the auditor's function is to make sure that all the requirements have been completed and followed by the management and employees of the organization. How do they do that? By generating a checklist, so that they know how to go about evaluate their process(es). That checklist focuses on the specificity of the requirement, its effectiveness, and whether or not it meets the customer requirements. So, for example,

1. The purpose of life cycle perspective is to prioritize actions that can reduce environmental impacts. If we examine the supply chain which impacts (energy, raw materials, water, and logistics); if we look at the direct company impacts (manufacturing emissions, fleet emissions, landfill waste, and recycle rate); and evaluate downstream impacts (products in use, and product disposal), that covers typically the entire life cycle (supplier to customer). So, by examining the life cycle, the auditor may consider to ask questions that deal with upstream and downstream chain:
   a. What does the organization do to help with the identification, evaluation, and interpretation of the environmental aspects?
   b. How is significance measured?
   c. How processes and products are selected for evaluation?
   d. How is risk associated with threads and opportunities to ensure the desired or expected outcomes?
   e. How does the organization evaluate the risk to be reduced or prevented?
   f. How does the organization improve transparency between management and supply chain?
   g. How does the organization treat the unintended consequences?
2. The effects of uncertainty must be calculated for each risk (Risk = likelihood of occurrence × consequences of that occurrence – impact) and each environmental aspect for the organization's EMS – which should cover the following:
   a. The definition of the rating scale (how and why was it selected)
   b. The definition of each risk or threat and their consequences
   c. The specific definitions of the risk classification in order of priority

   d. The definition of the specific resources to the risk management plan on hand

   e. The precise definition of the risk management plan as it relates to the avoidance, minimization, and mitigation

It is also important to note that the ISO 14001:2015 can be used in whole or in part to systematically improve environmental management. Claims of conformity to ISO 14001:2015, however, are not acceptable unless all its requirements are incorporated into an organization's environmental management system and fulfilled without exclusion.

With all that background info, the following cursory checklist is generated. By no means, it is an exhausted list. Rather, it is given here to help the auditor generate their own with more specificity and clarity for the organization that is about to undertake the commitment to follow the ISO 14001.

THE STRUCTURE AND CONTENT OF THE STANDARD

Clause 4.1 and 4.2 Context, needs, and expectation of interested parties. Things to consider:

4.1 Understanding the organization and its context:

a. Determine external and internal issues relevant to its purpose and that affect the achievement of intended outcomes
   i. Affected by
   ii. Capable of affecting the organization.

4.2 Understanding and identifying the needs and expectations of interested parties

a. Interested parties
b. Needs, expectations, and/or requirements. Typical concerns may be in the areas of:
c. Has your organization
   i. Identified external issues (proactive actions) that could be affected by products, services, or activities performed by your organization?
   ii. Identified external issues (proactive actions) that could be capable of affecting your organization's ability to deliver products, services, or activities?
   iii. Identified internal issues (proactive actions) that could be affected by products, services, or activities performed by your organization?
   iv. Identified internal issues (proactive actions) that could be capable of affecting your organization's ability to deliver products, services, or activities?
   v. Identified interested parties relevant to the EMS?
   vi. Determined relevant needs and expectations of interested parties?
   vii. Provided process for input from internal and external interested parties?

4.3 Determining scope of environmental management

In Appendix A.4.3, one reads that "In setting the scope, the credibility of the EMS depends upon the choice of organizational boundaries. The organization must consider the extend of control or influence it can exert over activities, products, and services using a life cycle perspective. Scoping should not be used to exclude activities, products, services or facilities that have or can have significant environmental aspects or to evade its compliance obligations.

The scope is a factual and representative statement of the organization's operations included within its environmental management system boundaries that should not mislead interested parties. The organization is obligated to make the final scope statement available to interested parties."

So, the issues for the auditor are to ask relative questions that will validate the scope of the EMS for validating the scope. Therefore, some typical questions may be in the form of: Did your organization take into consideration the following?

a. External and internal issues as being identified in 4.1 context?
b. Compliance obligations as identified in 4.2?
c. Needs and expectations of interested parties?
d. Your organizational unit(s), function(s), and physical boundaries in (a) its activities, products and services, and (b) its authority and ability to exercise control and influence?
e. Whether scope takes into consideration life cycle perspective?
f. Whether it excludes activities, products, services, or facilities that have potential significant environmental aspects or evade compliance obligations? (If yes, that may present serious legal issues).

4.4 Environmental management system. Some things to consider in generating a checklist are as follows:

a. Has your organization established processes to achieve the desired environmental performance results?
b. Are your EMS requirements integrated into business processes, such as design, development, purchasing, human resources, sales, and marketing?
c. Does your EMS incorporate issues related to context of the organization?
d. Does your EMS incorporate issues related to interested parties?

Clause 5.1 Leadership and commitment. Things to consider:

a. Has your organization clearly identified top management in your organization?
b. Has top management been briefed on the overall requirements of the ISO 14001?
c. Have the EMS-specific responsibilities which top management should be personally involved or should direct been clearly communicated?
d. Does top management understand they may delegate responsibility to others, but must retain accountability for ensuring the actions are performed?

e. Is top management committed to demonstrate leadership and support of EMS in the entire organization?

Clause 5.3 Leadership – organizational role, responsibilities, and authorities: Things to consider:

a. Has top management ensured
   i. Are responsibilities and authorities for relevant roles assigned?
   ii. Are responsibilities and authorities communicated within the organization?
   iii. Are responsibilities ensuring EMS conformance to ISO 14001 assigned?
   iv. Are responsibilities for reporting on performance of EMS to top management assigned?
   v. Is there a planned mechanism to report environmental performance to top management?

Clause 6.1.1 Planning, general, determine risks, and opportunities – Things to consider:

a. Have you identified environmental aspects, compliance, organizational context, and interested parties?
b. Have you identified risks throughout the life cycle of your products, activities, or services?
c. Have you ranked risks with quantifiable measure to identify significant risks associated with environmental aspects?
d. Have you identified options and/or alternatives to prevent or reduce undesired effects?
e. Do you have processes in place to address risks and emergencies?
f. Do you have confidence in your process or system that the planned activities will be carried out as planned? How do you assure this confidence? What mechanisms do you have or plan to control or mitigate the risks and emergencies?
g. Have you identified environmental aspects of products, activities, and services that an organization controls and influences?
h. Have you identified associated environmental impacts of products, activities, and services?
i. Have you considered life cycle perspective with respect to (a) supply chain, (b) product use, (c) end-of-life treatment or disposal, (d) purchased goods and services, and (e) maintain documented information regarding environmental aspects and environmental impacts?

Clause 6.2.1 Environmental objectives and planning to achieve. Things to consider: Has your organization established environmental objectives that

a. Are integrated into your organization's business process to support actions to achieve environmental objectives?

    b. Establish relevant functions and levels and take into account significant aspects?

    c. Reflect compliance obligations and consider risks and opportunities?

    d. Are consistent with environmental policy? Is it measurable, monitored, and communicated?

    e. Are they documented? Updated and appropriate?

    f. Have an action plan to achieve environmental objectives that states

        i. What will be done and what resources are required?

        ii. Who will be responsible? When will be completed?

       iii. How will the results be evaluated?

Clause 7 Support: Resources, competence, and awareness. Things to consider: Has the organization

    a. Determined and provided resources needed to establish, implement, maintain, and ensure continual improvement for EMS?

    b. Is there a process or a system in place to ensure that person(s) doing work under organization's control that affects it environmental performance and ability to fulfill compliance obligations are competent with respect to the standard?

    c. Determined training needs associated with environmental aspects and EMS specific to the standard?

    d. Established a system for documented information as evidence of competence?

    e. Established a system to ensure persons doing work under organization's control are aware of environmental policy, significant environmental aspects, potential environmental impacts, their contribution to EMS, and implications of not conforming?

Clause 7 Support: Communication (internal and external). Things to consider: Does your organization have

    a. A process for internal and external communication?

    b. Does the communication process cover what, when, with whom, and how information will be communicated?

    c. Communication protocol consider compliance obligations? Control to contribute to continual improvement?

    d. Does the system ensure environmental information communicated is consistent with information generated with EMS and is reliable?

    e. Retention procedure for communication?

    f. Is EMS information communicated among various levels and functions?

    g. Is there a communication process to enable persons doing work under organizations control to contribute to continual improvement?

Clause 7 Support: Documented information. Things to consider. Does your organization have?

a. System in place for consistently creating and updating documented information consistent with the ISO standard?

b. Documented info that is available and suitable for use, where and when it is needed?

c. A system to ensure documented information is adequately protected (improper use, integrity, confidentiality, even trust)?

d. A system for distribution, access, retrieval, use, storage, preservation, version control, retention, and disposition of documented information?

Clause 8 Planning and control: Things to consider. Does your organization have?

a. Operation criteria and control of processes, and a process to manage and control planned changes?

b. Does process incorporate review of consequence from intended changes, taking action to mitigate any adverse environmental effect?

c. A system to ensure outsourced processes are controlled or influenced and defined within the environmental management system?

d. System to review operational plans and changes from a life cycle perspective?

   i. How does your organization ensure environmental requirements are addressed in the D&D process for product or service for each life cycle stage?

   ii. How does your organization determine and document environmental requirements for procurement of products and service?

   iii. How does your organization communicate environmental requirements to external providers and contractors?

   iv. Provide information about potential significant environmental impacts associated with the transportation, delivery, use, end-of-life treatment, and final disposal of products and services.

Clause 8.2 Operation: Emergency preparedness and response. Things to consider. Does your organization have?

a. Processes established, implemented, controlled, and maintained to prepare for and respond to potential emergency situations?

b. Prepared plans to prevent or mitigate adverse environmental impacts from emergency situations?

c. Implement preventative or mitigation actions to avoid or minimize environmental consequences of emergency situations?

d. Periodical emergency tests for planned response actions? Periodically review and revise the emergency processes?

e. Provide relevant information and/or training related to emergency preparedness and response to relevant interested parties?

Clause 9.1 Performance evaluation; monitoring, measurement, analysis, and evaluation. Things to consider. Does your organization have documented information to demonstrate that your organization?

a. Monitors, measures, and evaluates its environmental performance?
b. Identifies what needs to be done as far as monitoring and measuring are concerned?
c. Established methods for monitoring and measuring, analyzing, and evaluating to ensure valid results?
d. Established criteria against which organization will evaluate environmental performance and appropriate indicators?
e. Stated when monitoring and measuring will occur (frequency)?
f. Identify when results will be analyzed and evaluated?
g. Evaluate its environmental performance (equipment, tools, etc.)?
h. Calibrate and maintain equipment? Is this process documented? How often does it place?
i. Evaluate the effectiveness of management system?
j. Communicate environmental performance to internal and external stakeholders? (See Table 5.2.)
k. Retain documented information as evidence of monitoring, measuring, analyzing, and evaluating results?

Clause 9.1.2 Evaluation and compliance. Things to consider. Does your organization?

a. Have a process to evaluate fulfillment of compliance obligations?
b. Have predetermined frequency that compliance will be evaluated?
c. Have method(s) to evaluate compliance and take action, if needed?
d. Have a system or process to maintain knowledge (database for "things learned" with retrievability ability) and understanding of its compliance status?
e. Retain documents as evidence of compliance evaluation results?

Clause 9.2 Performance evaluation: Internal audit. Things to consider. Does your organization have?

a. Established an internal audit program for the newest ISO 14001?
b. Identified frequency (weekly, monthly, quarterly, or yearly), method (process, product, service, layered), responsibilities, planning requirements, and reporting?

TABLE 5.2
Interested Stakeholders

Internal Interested Parties	External Interested Parties
Board of directors	Legislators
Top management	Regulators
Management team	Special interest groups
Union representation	Shareholders
Suppliers	Consumers
Employees	Community
Contractors	

c. Identified audit criteria and scope for each audit?

d. Do auditors conduct audits (they MUST) to ensure objectivity and impartiality?

e. Are the auditors trained appropriately? Are they certified? If not, what qualifications do they present so that they can conduct the audit?

f. Report internal audit results to management?

g. Do they have the appropriate and applicable documentation as evidence to support both the audit system and/or the findings?

h. Is top management (plant manager) part of the review team at predetermined intervals and/or after each audit? A typical review includes but not limited to

   i. Status of actions from previous management reviews?

  ii. Changes and adequacy of resources?

 iii. Report of environmental objectives achievement?

 iv. Organization's environmental performance?

  v. Communications from interested parties?

 vi. Opportunities for improvements? What stops us from adapting some of them on apriority basis?

ISO 18001

As we mentioned in Chapter 1, OHSAS 18001, Occupational Health and Safety Assessment Series (officially BS OHSAS 18001), is a British Standard for occupational health and safety management systems. Compliance with it enables organizations to demonstrate that they had a system in place for occupational health and safety. BSI is phasing out this standard and is adopting the ISO 45001 as BS ISO 45001 by March 2021 with very few changes and/or clarifications. Therefore, since this standard probably will be expired by the time this work is published, we do not provide a specific checklist. However, the checklist that follows covers all the requirements of OH&S and may be used accordingly.

ISO 45001:2018

ISO 45001:2018 has replaced the ISO 18001, and it is intended for use by organizations seeking to manage their environmental responsibilities in a systematic manner that facilitates the environmental principle of sustainability. To keep this consistency and focus of the standard, we suggest that the following five points be consider with extra attention since they are different from the ISO 18001.

- Context of the organization (Clause 4.1): The organization shall determine internal and external issues that are relevant to its purpose and that affect its ability to achieve the intended outcome(s) of its OH&S management system.
- Understanding the needs and expectations of workers and other interested parties (Clause 4.2): interested parties are workers, suppliers, subcontractors, clients, and regulatory authorities.
- Risk and opportunities (Clauses 6.1.1, 6.1.2.3, 6.1.4): companies are to determine, consider, and, where necessary, take action to address any risks or

opportunities that may impact (either positively or negatively) the ability of the management system to deliver its intended results, including enhanced health and safety at the workplace.
- Leadership and management commitment (Clause 5.1) have stronger emphasis on top management to actively engage and take accountability for the effectiveness of the management system.
- Planning: (Clause 6)

CLAUSE 4: CONTEXT OF THE ORGANIZATION

4.1 Understanding the organization and its context

a. Have you determined external and internal issues that are relevant to your purpose and your strategic direction and that affect your ability to achieve the intended outcomes of your Occupational Health and Safety management system?
b. How do you monitor and review information about these external and internal issues?

4.2 Understanding the needs and expectations of workers and other interested parties: Have you determined the following?

a. The interested parties in addition to workers that are relevant to the Occupational Health and Safety Management system?
b. The needs and expectations of these interested parties that are relevant to the Occupational Health and Safety Management System?
c. Which of these needs and expectations are or could become legal requirements and other requirements?
d. How do you monitor and review information about these interested parties and their relevant needs and expectations?

4.3 Determining the scope of the OH&S management system:

a. Have you determined the boundaries and applicability of the OH&S management system to establish your scope?
b. When determining the scope of the OH&S management system how did you consider:
   i.   The external and internal issues referred to in 4.1?
   ii. The requirements of relevant interested parties referred to in 4.2?
c. Take into account the planned or performed work-related activities?
d. Is the scope available as documented information?

4.4 OH&S Management System:
Have you implemented and have the system in place to maintain and continually improve your OH&S management system, including the processes needed and their interactions, in accordance with the requirements of ISO 45001?

CLAUSE 5: LEADERSHIP

5.1 Leadership and commitment: How does top management demonstrate leadership and commitment with respect to the OH&S management system?

a. Taking overall responsibility and accountability for the prevention of work-related injury and ill health, as well as the provision of safe and healthy workplaces and activities?
b. Ensuring that the OH&S policy and related OH&S objectives are established for the OH&S management system and are compatible with the strategic direction of the organization?
c. Ensuring the integration of the OH&S management system requirements into the organization's business processes?
d. Ensuring that the resources needed for the OH&S management system are available?
e. Communicating the importance of effective OH&S management and conforming to the OH&S management system requirements?
f. Ensuring that the OH&S management system achieves its intended outcomes?
g. Directing and supporting workers to contribute to the effectiveness of the OH&S management system?
h. Ensuring and promoting continual improvement?
i. Supporting other relevant management roles to demonstrate their leadership as it applies to their areas of responsibility?
j. Developing, leading, and promoting a culture in the organization that supports the intended outcomes of the OH&S management system?
k. Protecting workers from reprisals when reporting incidents, hazards, risks, and opportunities?
l. Ensuring the organization establishes and implements a process(es) for consultation and participation of workers?
m. Supporting the establishment and functioning of health and safety committee?

5.2 OH&S Policy: Have top management established, implemented, and maintained an OH&S policy that

a. Includes a commitment to provide safe and healthy working conditions for the prevention of work-related injury and ill health and is appropriate to the purpose, size, and context of the organization and to the specific nature of its OH&S risks and opportunities?
b. Provides a framework for setting OH&S objectives?
c. Includes a commitment to fulfill legal requirements and other requirements?
d. Includes a commitment to eliminate hazards and reduce OH&S risks?
e. Includes commitment to continual improvement of the OH&S management system?
f. Includes a commitment to consultation and participation of workers and where they exist workers representative?

g. Is the OH&S policy
   i. Available as documented information?
   ii. Communicated within the organization?
   iii. Available to interested parties?
   iv. Relevant and appropriate?

5.3 Organizational roles, responsibilities, and authorities:

a. Does top management ensure that the responsibilities and authorities for relevant roles within the OH&S management system are assigned, available as documented information, communicated, and understood at all levels within the organization?
b. Do workers assume responsibility for those aspects of the OH&S management system for which they have control?
c. Has top management assigned the responsibility and authority for
   i. Ensuring that the OH&S management system conforms to the requirements of ISO 45001?
   ii. Reporting on the performance of the OH&S management system to top management?

5.4 Consultation and of workers: Has your organization established, implemented, and maintained a process(es) for consultation and participation of workers at all applicable levels and functions, and where they exist, workers representatives, in the development, performance evaluation, and actions for improvement of the OH&S system? Does the organization

a. Provide mechanisms, time, training, and resources necessary for consultation and participation?
b. Provide timely access to clear, understandable, and relevant information about the OS&H management system?
c. Determine and remove obstacles or barriers to participation and minimize those that cannot be removed?
d. Emphasize the consultation of non-managerial workers on the following:
   i. Determining the needs and expectations of interested parties?
   ii. Establishing the OH&S policy?
   iii. Assigning organizational roles, responsibilities, and authorities, as applicable?
   iv. Determining how to fulfill legal and other requirements?
   v. Establish and plan to achieve OH&S objectives?
   vi. Determining applicable controls for outsourcing, procurement, and contractors?
   vii. Determining what needs to be monitored, measured, and evaluated?
   viii. Planning, establishing, implementing, and maintaining an audit program?
   ix. Ensuring continual improvement?

e. Emphasize participation of non-managerial workers in the following:
   i. Determining the mechanisms for their consultation and participation?
   ii. Identifying hazards and assessing risks and opportunities?
   iii. Determining actions to eliminate hazards and reduce OH&S risks?
   iv. Determining competence requirements, training needs, training, and evaluating training?
   v. Determining what needs to be communicated and how it is to be done?
   vi. Determining control measures and their effective implementation and use?
   vii. Investing incidents and nonconformities and determine corrective action?

Clause 6: Planning

6.1 Actions to address risks and opportunities

6.1.1 General: When planning for the OH&S management system, have you considered the issues referred to in 4.1 and the requirements referred to in 4.2 and 4.3 and determined the risks and opportunities that need to be addressed to

a. Give assurance that the OH&S management system can achieve its intended outcomes?
b. Prevent, or reduce, undesired effects?
c. Achieve continual improvement?
d. When determining the risks and opportunities for the OH&S management system and its intended outcome, Has the organization taken into account
   • Hazards?
   • OH&S risks and other risks?
   • OH&S opportunities and other opportunities?
   • Legal and other requirements?
e. Has your organization in its planning process determined and assessed the risks and opportunities relevant to the intended outcomes of the OH&S system associated with planned changes permanent or temporary before the change is implemented?
f. Does your organization maintain documented information on
   i. Risks and opportunities?
   ii. The process and actions needed to determine and address its risks and opportunities to the extent necessary to have confidence that they are carried out as planned?

6.1.2 Hazards identification and assessment of risks and opportunities

6.1.2.1 Hazard identification: Has the organization established, implemented, and maintained a process(es) for hazard identification that is ongoing and proactive? Do the processes take into account, but not be limited to,

a. How work is organized, social factors (including workload, work hours, victimization, harassment, and bullying), leadership, and the culture of the organization?

  b. Routine and non-routine activities and situations, including hazards arising from
     1. Infrastructure, equipment, materials, substances, and the physical conditions of the workplace?
     2. Product and service design, research, development, testing, production, assembly, construction, service delivery, maintenance, and disposal?
     3. Human factors? How work is performed?
  c. Past relevant incidents, internal or external to the organization, including emergencies, and their causes?
  d. Potential emergency situations?
  e. People, including consideration of,
     1. Those with access to the workplace and their activities, including workers, contractors, visitors and other persons?
     2. Those in the vicinity of the workplace who can be affected by the activities of the organization?
     3. Workers at a location not under the direct control of the organization?
  f. Other issues, including consideration of,
     1. The design of work areas, processes, installations, machinery/equipment, operating procedures, and work organization, including their adaptation to the needs and capabilities of the workers involved?
     2. Situations occurring in the vicinity of the workplace caused by work-related activities under the control of the organization?
     3. Situations not controlled by the organization and occurring in the vicinity of the workplace that can cause injury and ill health to persons in the workplace?
  g. Actual or proposed changes in organization, operations, processes, activities, and the OH&S management system?
  h. Changes in knowledge of, and information about, hazards?

6.1.2.2 Assessment of OH&S risks and other risks to the OH&S management system: Has the organization established implemented and maintained a process to

  a. Assess OH&S risks from the identified hazards, while taking into account the effectiveness of existing controls?
  b. Determine and assess the other risks related to the establishment, implementation, operation, and maintenance of the OH&S management system?
  c. Has the organization's methodologies and criteria for the assessment of OH&S risks been defined with respect to the scope, nature, and timing to ensure they are proactive rather than reactive and are used in a systematic way?
  d. Does the organization maintain and retain documented information on the methodologies and criteria?

6.1.2.3 Assessment of OH&S opportunities and other opportunities for the OH&S management system: Has the organization established, implemented, and maintained processes to assess?

a. OH&S opportunities to enhance OH&S performance, while taking into account planned changes to the organization, its policies, its processes, and its activities and
   1. Opportunities to adapt work, work organization, and work environment to workers?
   2. Opportunities to eliminate hazards and reduce OH&S risks?
b. Other opportunities for improving the OH&S system?

6.1.3 Determination of legal requirements and other requirements: Has the organization established, implemented, and maintained processes to

a. Determine and have access to up to date legal requirements and other requirements that are applicable to the hazards, OH&S risks, and OH&S management system?
b. Determine how these legal requirements and other requirements apply to the organization and what needs to be communicated?
c. Take legal and other requirements into account when establishing implementing, maintaining, and continually improving its OH&S management system?
d. Does the organization maintain and retain information on its legal and other requirements?
e. How does the organization ensure its legal requirements are up to date and reflect any changes?

6.1.4 Planning Action: Does the organizations plan include?

a. Actions to address these risks and opportunities, address legal and other requirements and prepare for and respond to emergency situations?
b. How to integrate and implement the actions into its OH&S management system processes or other business processes?
c. Has the organization taken into account the hierarchy of controls and outputs and outputs from OH&S management system when planning to take action?
d. Does the organization take into account best practice, technological options, and financial, operational, and business requirements when planning its actions?

6.2 OH&S objectives and planning to achieve them
   6.2.1 Has your organization established OH&S objectives at relevant functions and levels that are needed to maintain and continually improve the OH&S management system? Are the OH&S objectives

a. Consistent with the OH&S policy?
b. Measurable or capable of performance evaluation?
c. Taking into account applicable requirements, the results of the assessment of risks and opportunities and the results of consultation with worker and workers representatives?

    d. Monitored? Communicated? Updated as appropriate?

    e. Do you maintain and retain documented information on the OH&S objectives?

6.2.2 When planning how to achieve your OH&S objectives, has your organization determined?

    a. What will be done?

    b. What resources will be required?

    c. Who will be responsible?

    d. When it will be completed?

    e. How the results will be evaluated including indicators for monitoring?

    f. How the actions to achieve OH&S objectives will be integrated into the organizations business processes?

    g. Do you maintain and retain documented information on the OH&S plans?

CLAUSE 7: SUPPORT

7.1 Resources: Has your organization determined and provided the resources needed for the establishment, implementation, maintenance, and continual improvement of the OH&S management system?

    7.2 Competence: Has your organization

    a. Determined the necessary competence of workers that affects the performance and effectiveness of the OH&S management system?

    b. Ensured that these workers are competent (including the ability to identify hazards) on the basis of appropriate education, training, or experience?

    c. Where applicable, taken actions to acquire and maintain the necessary competence, and evaluated the effectiveness of the actions taken?

    d. Retained appropriate documented information as evidence of competence?

7.3 Awareness: How does the organization ensure that workers are aware of?

    a. The OH&S and objectives policy?

    b. Their contribution to the effectiveness of the OH&S system including the benefits of improved OH&S performance?

    c. The implications of not conforming to the OH&S management system requirements?

    d. Incidents and the outcomes of investigations that are relevant to them?

    e. Hazards, OH&S risks, and actions determined that are relevant to them?

    f. The ability to remove themselves from work situations that they consider presents an imminent and serious danger to their life or health, as well as the arrangements for protecting them from undue consequences for doing so?

7.4 Communication

    7.4.1 General: How have you determined the internal and external communications relevant to the OH&S management system, including

a. On what it will communicate? When to communicate?
b. With whom to communicate:
    1. Internally among the various levels and functions of the organization?
    2. Among contractors and visitors to the workplace?
    3. Among other interested parties?
c. How to communicate? How does the organization take into account diversity (gender, language, culture, literacy, disability) aspects when considering communication needs?
d. How are the views of interested parties considered in establishing communication processes? In establishing communication processes, has legal and other requirements been taken into account and that the information is consistent with other information generated from the system and reliable?
e. Who responds to relevant communications on its OH&S management system?
f. In what form is documented information retained as evidence of communications?

7.4.2 Internal communication: Has the organization ensured that

a. Internally communicated information is relevant to the OH&S management system among various levels and functions of the organization. Does it include changes to the OH&S management system?
b. Workers are able to contribute to continual improvement?

7.4.3 External communication

a. Has the company have an external communication process?
b. How does external communication of OH&S information take into account legal and other requirements?

7.5 Documented information
7.5.1 Does your organization's OH&S management system include

a. Documented information required by ISO 45001?
b. Documented information determined by the organization as being necessary for the effectiveness of the OH&S management system?

7.5.2 When creating and updating documented information, how does your organization ensure appropriate

a. Identification and description (e.g., a title, date, author, or reference number)?
b. Format (e.g., language, software version, graphics) and media (e.g., paper, electronic)?
c. Review and approval for suitability and adequacy?

7.5.3 How do you ensure documented information required by your OH&S management system and by ISO 45001 is controlled to ensure?

    a. It is available and suitable for use, where and when it is needed?
    b. It is adequately protected (e.g., from loss of confidentiality, improper use, or loss of integrity)?

7.5.3.2 For the control of documented information, how does your organization address the following activities, as applicable:

    a. Distribution, access, retrieval, and use?
    b. Storage and preservation, including preservation of legibility?
    c. Control of changes (e.g., version control)?
    d. Retention and disposition?
    e. How do you ensure documented information of external origin is identified and controlled?

CLAUSE 8: OPERATION

8.1 Operational planning and control

8.1.1 General: Does your organization plan, implement, and control the processes (see 4.4) needed to meet the requirements of the OH&S management system and to implement the actions determined in Clause 6 by

    a. Establishing criteria for the processes?
    b. Implementing control of the processes in accordance with the criteria?
    c. Maintaining and keeping documented information to the extent necessary to have confidence that processes are being carried out as planned?
    d. Adapting to workers?
        i. How does your organization coordinate the relevant parts of OH&S management system with other organizations in multi-employer situations?
        ii. How does your organization ensure that outsourced processes are controlled (see 8.4)?

8.1.2 Eliminating hazards and reducing OH&S risks

Has the organization established, implemented, and maintained processes for the elimination of hazards and reduction of OH&S risks using the following hierarchy of controls:

    a. Eliminate the hazard?
    b. Substitute with less hazardous process, operations, materials, or equipment?
    c. Use engineering controls and reorganization of work?
    d. Use administration controls, including training?
    e. Use adequate personal protective equipment?

8.1.3 Management of change
Has the organization established processes for the implementation and control of planned temporary and permanent changes that impact performance, including

    a. New products, services, and processes, or changes to existing products, services, and processes, including
       i. Workplace locations and surroundings?
      ii. Working organization?
     iii. Working conditions?
     iv. Equipment?
      v. Work force?
    b. Changes to legal requirements and other requirements?
    c. Changes to knowledge or information about hazards and OH&S risks?
    d. Developments in knowledge and technology? (Does the organization review the consequences of unintended changes, taking action to mitigate any adverse effects, as necessary?)

8.1.4 Procurement
8.1.4.1
Has the organization established, implemented, and maintained processes to control the procurement of products and services in order to ensure their conformity to its OH&S management system?
8.1.4.2
Does the organization coordinate its procurement processes with its contractors, in order to identify hazards and assess and control the OH&S risks arising from

    a. The contractors' activities and operations that impact the organization?
    b. The organization's activities and operations that impact the contractors' workers?
    c. The contractors' activities and operations that impact other interested parties in the workplace?
    d. How does the organization ensure that the requirements of its OH&S management system are met by contractors and their workers?
    e. Do the organization's procurement processes define and apply occupational health and safety criteria for the selection of contractors?

8.1.4.3

    a. How does the organization ensure outsourced functions and processes are controlled?
    b. Does the organization ensure that its outsourcing arrangements are consistent with legal requirements and other requirements and with achieving the intended outcomes of the OH&S management system?
    c. Has the type and degree of control to be applied to these functions and processes been defined within the OH&S management system?

8.2 Emergency preparation and response

Has the organization established, implemented, and maintained the processes needed to prepare for and respond to potential emergency situations identified in 6.1.2.1 and do they include

a. Establishing a planned response to emergency situations including provision of first aid?
b. Providing training for the planned response?
c. Periodically testing and exercising the planned response capability?
d. Evaluating performance and as necessary, revising the planned response, including after testing and in particular after the occurrence of an emergency situation?
e. Communicating and providing relevant information to all workers on their duties and responsibilities?
f. Communicating relevant information to contractors, visitors, emergency response services, government authorities, and as appropriate local community?
g. Taking into account the needs and capabilities of all relevant interested parties and ensuring their involvement, as appropriate, in the development of the planned response?
h. Has the organization maintained documented information on the process and on the plans for responding to potential emergency situations?

Clause 9: Performance Evaluation

9.1 Monitoring, measurement, analysis, and evaluation

9.1.1 General: The organization shall establish, implement, and maintain processes for monitoring, measurement analysis, and performance evaluation. How does your organization determine?

a. What needs to be monitored and measured:
    1. The extent to which legal requirements and other requirements are met?
    2. Its activities and operations related to identified hazards, risks, and opportunities?
    3. Progress towards achieving OH&S objective?
    4. Effectiveness of operational and other controls?
b. The methods for monitoring, measurement, analysis, and performance evaluation needed to ensure valid results?
c. The criteria against which the organization will evaluate its OH&S performance?
d. When the monitoring and measuring shall be performed?
e. When the results from monitoring and measurement shall be analyzed and evaluated and communicated?
f. How does your organization evaluate the performance and the effectiveness of the OH&S management system?

g. How does the organization ensure that monitoring and measuring equipment is calibrated or verified as applicable, and used and maintained as appropriate?

h. In what form does your organization retain appropriate documented information as evidence of the monitoring, measurement, analysis and performance evaluation, and maintenance, calibration, or verification of measuring equipment?

9.1.2 Evaluation of compliance: How does your organization establish implement and maintain processes for evaluating compliance with legal and other requirements? Does the evaluation include

a. Determining the frequency and method(s) for the evaluation of compliance?
b. Evaluating compliance and taking action if needed?
c. Maintaining knowledge and understanding of its compliance status with legal requirements and other requirements?
d. Retaining documented information of the compliance evaluation results?

9.2 Internal audit: Does your organization conduct internal audits at planned intervals to provide information on whether the OH&S management system:

a. Conforms to
  1. The organization's own requirements for its OH&S management system, including policy and objectives?
  2. The requirements of this international standard?
b. Is effectively implemented and maintained?

9.2.2 Internal audit program: Does your organization

a. Plan, establish, implement, and maintain an audit program(s) including the frequency, methods, responsibilities, planning requirements, and reporting, which shall take into consideration the importance of the processes concerned, and the results of previous audits?
b. Define the audit criteria and scope for each audit?
c. Select auditors and conduct audits to ensure objectivity and the impartiality of the audit process?
d. Ensure that the results of the audits are reported to relevant management; and ensure results of internal audits are reported to workers and where they exist, workers representatives, and other relevant interested parties?
e. Take action to address nonconformity and continually improve its OH&S audit program and the audit results?
f. Retain documented information as evidence of the implementation of the audit program and the audit results?

9.3 Management review: ISO 45001 requires "Top management shall review the organization's OH&S management system, at planned intervals, to ensure its continuing suitability, adequacy, effectiveness." What format does this review take? Is your organization's management review planned and carried out taking into consideration?

a. The status of actions from previous management reviews?
b. Changes in external and internal issues that are relevant to the OH&S management system including
    1. Needs and expectations of interested parties?
    2. Legal requirements and other requirements?
    3. Risks and opportunities?
c. The extent to which OH&S policy and objectives have been met?
d. Information on the OH&S performance, including
    1. Incidents nonconformities and corrective actions and continual improvement?
    2. Monitoring and measurement results?
    3. Results of evaluation of compliance with legal requirements other requirements?
    4. Audit results?
    5. Consultation and participation of workers?
    6. Risks and opportunities?
e. Adequacy of resources for maintaining an effective OH&S system?
f. Relevant communication with interested parties?
g. Opportunities for continual improvement?
h. Do the outputs of the management review include decisions and actions related to
    1. The continuing suitability, adequacy, and effectiveness in achieving the intended outcomes?
    2. Continual improvement opportunities?
    3. Any need for changes to the OH&S management system?
    4. Resource needs?
    5. Actions needed?
    6. Opportunities to improve integration of the OH&S system with other business processes?
    7. Any implications for the strategic direction of the organization?
i. How are the relevant outputs from management review communicated to workers and where they exist workers representatives?
j. In what form does your organization retain documented information as evidence of the results of management reviews?

CLAUSE 10: IMPROVEMENT

10.1 General: How do you determine and select opportunities for improvement and implement any necessary actions to achieve intended outcomes of your OH&S management system?

10.2 Incident, nonconformity, and corrective action: When an incident or nonconformity occurs, how does your organization

   a. React in a timely manner to the incident or nonconformity and, as applicable,
     1. Take action to control and correct it?
     2. Deal with the consequences?
   b. Evaluate, with the participation of workers and the involvement of other relevant interested parties, the need for corrective action to eliminate the root cause(s) of the incident or nonconformity, in order that it does not recur or occur elsewhere, by
     1. Investigating the incident or reviewing the nonconformity?
     2. Determining the causes of the incident or nonconformity?
     3. Determining if similar incidents have occurred, if nonconformities exist, or if could potentially occur?
   c. Review existing assessments of OH&S risks and other risks, as appropriate?
   d. Determine and implement any action needed, including corrective action, in accordance with the hierarchy of controls and the management of change?
   e. Assess OH&S risks that relate to new or changed hazards, prior to taking action?
   f. Review the effectiveness of any action taken, including corrective action?
   g. Make changes to the OH&S management system if necessary?
   h. Does your organization take corrective actions appropriate to the effects or potential effects of the incidents or nonconformities encountered?
     1. In what form does your organization retain documented information evidence of
       i. The nature of the incidents or nonconformities and any subsequent actions taken?
       ii. The results of any action and corrective action including their effectiveness?
   i. How is this information communicated to relevant workers, and, where applicable, workers representatives and other interested parties?

10.3 Continual improvement: How does your organization continually improve the suitability, adequacy, and effectiveness of the OH&S management system? How does your organization:

   a. Enhance OH&S performance?
   b. Promote a culture that supports the OH&S management system?
   c. Promote the participation of workers in implementing actions for continual improvement of the OH&S management system?
   d. Communicating the results of continual improvement workers and if appropriate workers representatives?
   e. Maintain and retain documented information as evidence of continual improvement?

Again, the auditor's function is to make sure that all these requirements have been completed and followed by the management and employees of the organization. How do they do that? By generating a checklist. Some typical generic questions have been identified in this section to make the process easier and functional. The reader will notice that the checklist focuses on the specificity of the requirement, its effectiveness, and whether or not it meets the customer requirements.

IATF 16949:2016

The following checklist can be used for both internal audit and gap analysis tools. The checklist given below has the requirements as given in standard IATF 16949:2016 and has to be used along with the requirements as given in Standard ISO 9001:2015. In addition, the user of this checklist must be aware that the IATF has its own checklist and for certification purposes, it must be used. The list provided here is for a guide for those organizations that want to do an internal evaluation of their own.

CLAUSE 4: CONTEXT OF THE ORGANIZATION

4.3 Determining the scope of the quality management system
    4.3.1 Determining the scope of the quality management system-supplemental

1. Are supporting functions, whether on-site or remote (such as design centers, corporate headquarters, and distribution centers), included in the scope of the Quality Management System (QMS)?
2. Have you taken the only permitted exclusion for this Automotive QMS Standard relates to the product D&D requirements within ISO 9001, Section 8.3 (Design and development of product and services)? Is the exclusion justified and maintained as documented information? Please note permitted exclusions do not include manufacturing process design

4.3.2 Customer-specific requirements

1. Is customer-specific requirements evaluated and included in the scope of the organization's quality management system?

4.4 Quality management system and its processes
    4.4.1.1 Conformance of product and processes

1. Has the organization ensured conformance of all products and processes, including service parts and those that are outsourced, to all applicable customer, statutory, and regulatory requirements?

4.4.1.2 Product safety

1. Does the organization have documented processes for the management of product safety-related products and manufacturing processes?
2. Does the organization have documented processes for identification of statutory and regulatory product safety requirements?

3. Does the organization have documented processes for customer notification of requirements in identification of statutory and regulatory product safety requirements?

4. Does the organization have documented processes for special approvals for design FMEA? Note: Special approval is an additional approval by the function (typically the customer) that is responsible to approve such documents with safety-related content.

5. Does the organization have documented processes for identification of product safety-related characteristics?

6. Does the organization have documented processes for identification and controls of safety-related characteristics of product and at the point of manufacture?

7. Does the organization have documented processes for special approval of control plans and process FMEAs?

8. Does the organization have documented processes for reaction plans?

9. Does the organization have documented processes for defined responsibilities, definition of escalation process and flow of information, including top management, and customer notification?

10. Does the organization have documented processes for training identified by the organization or customer for personnel involved in product safety-related products and associated manufacturing processes?

11. Does the organization have documented processes for changes of product or process shall be approved prior to implementation, including evaluation of potential effects on product safety from process and product changes?

12. Does the organization have documented processes for transfer of requirements with regard to product safety throughout the supply chain, including customer-designated sources?

13. Does the organization have documented processes for product traceability by manufactured lot (at a minimum) throughout the supply chain?

14. Does the organization have documented processes for lessons learned for new product introduction?

15. Does the organization have documented processes for training identified by the organization or customer for personnel involved in product safety-related products and associated manufacturing processes?

16. Does the organization have documented processes for changes of product or process shall be approved prior to implementation, including evaluation of potential effects on product safety from process and product changes?

17. Does the organization have documented processes for transfer of requirements with regard to product safety throughout the supply chain, including customer-designated sources?

18. Does the organization have documented processes for product traceability by manufactured lot (at a minimum) throughout the supply chain?

19. Does the organization have documented processes for lessons learned for new product introduction?

CLAUSE 5: LEADERSHIP

Clause 5.1 Leadership and commitments
   Clause 5.1.1 General
   5.1.1.1 Leadership and commitment – corporate responsibility

   1. Has the organization defined and implemented corporate responsibility policies, including at a minimum an anti-bribery policy, an employee code of conduct, and an ethics escalation policy ("whistle-blowing policy")?

5.1.1.2 Process effectiveness and efficiency

   1. Has top management reviewed the product realization processes and support processes to evaluate and improve their effectiveness and efficiency? Are the results of the process review activities included as input to the management review?

5.1.1.3 Process owners

   1. Has top management identified process owners who are responsible for managing the organization's processes and related outputs?
   2. Do process owners understand their roles and are they competent to perform those roles?

5.3 Organizational roles, responsibilities, and authorities
   5.3.1 Organizational roles, responsibilities, and authorities – supplemental

   1. Has top management assigned personnel with the responsibility and authority to ensure that customer requirements are met?
   2. Have these assignments been documented?
   3. Does this include but is not limited to the selection of special characteristics, setting quality objectives and related training, corrective and preventive actions, product D&D, capacity analysis, logistics information, customer scorecards, and customer portals?

5.3.2 Responsibility and authority for product requirements and corrective actions

   1. Has top management ensured that personnel responsible for conformity to product requirements have the authority to stop shipment and stop production to correct quality problems?
   2. In case it is not possible to stop production immediately, has top management ensured that the affected batch is contained and shipment to the customer prevented?
   3. Has top management ensured that personnel with authority and responsibility for corrective action are promptly informed of products or processes that do not conform to requirements to ensure that nonconforming product is

not shipped to the customer and that all potential nonconforming product is identified and contained?

4. Has top management ensured that production operations across all shifts are staffed with personnel in charge of, or delegated responsibility for, ensuring conformity to product requirements?

5. Has top management ensured that personnel with authority and responsibility for corrective action are promptly informed of products or processes that do not conform to requirements to ensure that nonconforming product is not shipped to the customer and that all potential nonconforming product is identified and contained?

6. Has top management ensured that production operations across all shifts are staffed with personnel in charge of, or delegated responsibility for, ensuring conformity to product requirements?

CLAUSE 6: PLANNING

6.1 Action to address risks and opportunities
   6.1.2.1 Risk analysis

1. Has the organization included in its risk analysis, at a minimum, lessons learned from product recalls, product audits, field returns and repairs, complaints, scrap, and rework?

2. Has the organization retained documented information as evidence of the results of risk analysis??

6.1.2.2 Preventive actions

1. Has the organization determined and implemented action(s) to eliminate the causes of potential nonconformities in order to prevent their occurrence?

2. Are preventive actions appropriate to the severity of the potential issues?

3. Has the organization established a process to lessen the impact of negative effects of risk?

4. Has the organization established a process to determining potential nonconformities and their causes?

5. Has the organization established a process to evaluating the need for action to prevent occurrence of nonconformities?

6. Has the organization established a process to determining and implementing action needed?

7. Has the organization established a process to documented information of action taken?

8. Has the organization established a process to reviewing the effectiveness of the preventive action taken?

9. Has the organization established a process to utilizing lessons learned to prevent recurrence in similar processes?

10. Has the organization established a process to utilizing lessons learned to prevent recurrence in similar processes?

6.1.2.3 Contingency plan

1. Has the organization identified and evaluated internal and external risks to all manufacturing processes and infrastructure equipment essential to maintain production output and to ensure that customer requirements are met?
2. Has the organization defined contingency plans according to risk and impact to the customer?
3. Has the organization prepared contingency plans for continuity of supply in the event of key equipment failures, interruption from externally provided products, processes, and services, recurring natural disasters, fire, utility interruptions, labor shortages, or infrastructure disruptions?
4. Has the organization included, as a supplement to the contingency plans, a notification process to the customer and other interested parties for the extent and duration of any situation impacting customer operations?
5. Has the organization periodically tested the contingency plans for effectiveness (e.g., simulations, as appropriate)?
6. Has the organization conducted contingency plan reviews at a minimum annually using a multidisciplinary team including top management, and updated as required?
7. Has the organization documented the contingency plans and retained documented information describing any revisions, including the person who authorized the change?
8. Do the contingency plans include provisions to validate that the manufactured product continues to meet customer specifications after the restart of production following an emergency in which production was stopped and if the regular shutdown processes were not followed?

6.2.6.2 Quality objectives and planning to achieve them
  6.2.2.1 Quality objectives and planning to achieve them – supplemental

1. Has top management ensured that quality objectives to meet customer requirements are defined, established, and maintained for relevant functions, processes, and levels throughout the organization?
2. Are the results of the organization's review regarding interested parties and their relevant requirements considered when the organization establishes its annual (at a minimum) quality objectives and related performance targets (internal and external)?

CLAUSE 7: SUPPORT

7.1 Resources
  7.1.3 Infrastructure
  7.1.3.1 Plant, facility, and equipment planning

1. Has the organization used a multidisciplinary approach including risk identification and risk mitigation methods for developing and improving plant, facility, and equipment plans?

2. In designing plant layouts, has the organization optimized material flow, material handling, and value-added use of floor space including control of nonconforming product?
3. In designing plant layouts facilitated synchronous material flow?
4. Are methods developed and implemented to evaluate manufacturing feasibility for new product or new operations?
5. Do manufacturing feasibility assessments include capacity planning?
6. Are these methods also applicable for evaluating proposed changes to existing operations?
7. Has the organization maintained process effectiveness, including periodic re-evaluation relative to risk, to incorporate any changes made during process approval, control plan maintenance, and verification of job set-ups?
8. Are assessments of manufacturing feasibility and evaluation of capacity planning inputs to management reviews?
9. As applicable, do these requirements should include the application of lean manufacturing principles and apply to on-site supplier activities?

7.1.4 Environment for the operation of processes
7.1.4.1 Environment for the operation of processes – supplemental

1. Has the organization maintained its premises in a state of order, cleanliness, and repair that is consistent with the product and manufacturing process needs?

7.1.5 Monitoring and measuring resources
7.1.5.1 General
7.1.5.1.1 Measurement systems analysis

1. Have statistical studies been conducted to analyze the variation present in the results of each type of inspection, measurement, and test equipment system identified in the control plan?
2. Do the analytical methods and acceptance criteria used conform to those in reference manuals on measurement systems analysis? Other analytical methods and acceptance criteria may be used if approved by the customer.
3. Are records of customer acceptance of alternative methods retained along with results from alternative measurement systems analysis?
4. Is prioritization of MSA studies focused on critical or special product or process characteristics?

7.1.5.2 Measurement traceability
7.1.5.2.1 Calibration/verification records

1. Does the organization have a documented process for managing calibration/verification records?
2. Are records of the calibration/verification activity for all gauges and measuring and test equipment (including employee-owned equipment relevant for measuring, customer-owned equipment, or on-site supplier-owned

equipment) needed to provide evidence of conformity to internal requirements, legislative and regulatory requirements, and customer-defined
requirements retained?

3. Has the organization ensured that calibration/verification activities and
records include revisions following engineering changes that impact measurement systems?

4. Has the organization ensured that calibration/verification activities and
records include any out-of-specification readings as received for calibration/
verification?

5. Has the organization ensured that calibration/verification activities and
records include an assessment of the risk of the intended use of the product
caused by the out-of-specification condition?

6. Has the organization ensured that when a piece of inspection measurement
and test equipment is found to be out of calibration or defective during its
planned verification or calibration or during its use, documented information on the validity of previous measurement results obtained with this
piece of inspection measurement and test equipment is retained, including
the associated standard's last calibration date and the next due date on the
calibration report?

7. Has the organization ensured that notification is sent to the customer if suspect product or material has been shipped?

8. Has the organization ensured that calibration/verification activities and
records include statements of conformity to specification after calibration/
verification?

9. Has the organization ensured that calibration/verification activities and
records include verification that the software version used for product and
process controls is as specified?

10. Has the organization ensured that calibration/verification activities and
records include records of the calibration and maintenance activities for
all gauging including employee-owned equipment, customer-owned equipment, or on-site supplier-owned equipment?

11. Has the organization ensured that calibration/verification activities
and records include production-related software verification used for
product and process control including software installed on employee-
owned equipment, customer-owned equipment, or on-site supplier-owned
equipment?

7.1.5.3.1 Laboratory requirements: internal laboratory

1. Does the organization's internal laboratory facility have a defined scope that
includes its capability to perform the required inspection, test, or calibration
services?

2. Is this laboratory scope included in the quality management system
documentation?

3. Has the laboratory specified and implemented requirements for adequacy of
the laboratory technical procedures?

4. Has the laboratory specified and implemented requirements for competency of the laboratory personnel?
5. Has the laboratory specified and implemented requirements for testing of the product?
6. Does the laboratory have the capability to perform these services correctly, traceable to the relevant process standard such as ASTM and EN?
7. When no national or international standard(s) is available, has the organization defined and implemented a methodology to verify measurement system capability?
8. Has the laboratory specified and implemented requirements for customer requirements?
9. Has the laboratory specified and implemented requirements for review of the related records?
10. Does the laboratory have a third-party accreditation to ISO/IEC 17025 (or equivalent) to demonstrate the organization's in-house laboratory conformity to the abovementioned requirements?

7.1.5.3.2.1 Laboratory requirements: external laboratory

1. Do external/commercial/independent laboratory facilities used for inspection, test, or calibration services by the organization have a defined laboratory scope that includes the capability to perform the required inspection, test, or calibration?
2. Is the external laboratory accredited to ISO/IEC 17025 or national equivalent and includes the relevant inspection, test, or calibration service in the scope of the accreditation (certificate) where the certificate of calibration or test report includes the mark of a national accreditation body, or there is evidence that the external laboratory is acceptable to the customer?

Note: Such evidence may be demonstrated by customer assessment, for example, or by customer-approved second-party assessment that the laboratory meets the intent of ISO/IEC 17025 or national equivalent. The second-party assessment may be performed by the organization assessing the laboratory using a customer-approved method of assessment. Calibration services may be performed by the equipment manufacturer when a qualified laboratory is not available for a given piece of equipment. In such cases, the organization shall ensure that the requirements listed in Section 7.1.5.3.1 have been met. Use of calibration services, other than by qualified (or customer-accepted) laboratories, may be subject to government regulatory confirmation, if required.

7.2 Competence
7.2.1 Competence – supplemental

1. Has the organization established and maintained a documented process for identifying training needs including awareness and achieving competence of all personnel performing activities affecting conformity to product and process requirements?

2. Are personnel performing specific assigned tasks qualified, as required, with particular attention to the satisfaction of customer requirements?

7.2.2 Competence – on-the-job training

1. Does the organization provide on-the-job training, which includes customer requirements training, for personnel in any new or modified responsibilities affecting conformity to quality requirements, internal requirements, and regulatory or legislative requirements?
2. Does this include contract or agency personnel?
3. Is the level of detail required for on-the-job training commensurate with the level of education the personnel possess and the complexity of the task they are required to perform for their daily work?
4. Are persons whose work can affect quality informed about the consequences of nonconformity to customer requirements?

7.2.3 Internal auditor competency

1. Does the organization have a documented process to verify that internal auditors are competent, taking into account any customer-specific requirements?
2. Does the organization maintain a list of qualified internal auditors?
3. Are quality management system auditors, manufacturing process auditors, and product auditors all able to demonstrate the understanding of the automotive process approach for auditing, including risk-based thinking?
4. Are the auditors able to demonstrate the understanding of applicable customer-specific requirements?
5. Are the auditors able to demonstrate the understanding of applicable ISO 9001 and IATF 16949 requirements related to the scope of the audit?
6. Are the auditors able to demonstrate the understanding of applicable core tool requirements related to the scope of the audit?
7. Are the auditors able to demonstrate the understanding how to plan, conduct, report, and close out audit findings?
8. Do manufacturing process auditors demonstrate technical understanding of the relevant manufacturing process to be audited, including process risk analysis such as PFMEA and control plan?
9. Do product auditors demonstrate competence in understanding product requirements and use of relevant measuring and test equipment to verify product conformity?

Where training is provided to achieve competency, is documented information retained to demonstrate the trainer's competency with the above requirements?

10. Is maintenance of and improvement in internal auditor competence demonstrated through executing a minimum number of audits per year, as defined by the organization?
11. Is maintenance of and improvement in internal auditor competence demonstrated through maintaining knowledge of relevant requirements based on

internal changes (e.g., process technology, product technology) and external changes (e.g., ISO 9001, IATF 16949, core tools, and customer-specific requirements)?

7.2.4 Second-party auditor competency

1. Does the organization demonstrate the competence of the auditors undertaking the second-party audits?
2. Do second-party auditors meet customer-specific requirements for auditor qualification and demonstrate the understanding of the automotive process approach to auditing, including risk-based thinking?
3. Do second-party auditors demonstrate the understanding of applicable customer and organization-specific requirements?
4. Do second-party auditors demonstrate the understanding of applicable ISO 9001 and IATF 16949 requirements related to the scope of the audit?
5. Do second-party auditors demonstrate the understanding of applicable manufacturing process to be audited, including PFMEA and control plan?
6. Do second-party auditors demonstrate the understanding of applicable core tool requirements related to the scope of the audit?
7. Do second-party auditors demonstrate the understanding of how to plan, conduct, prepare audit reports, and close out audit findings?

7.3 Awareness
7.3.1 Awareness – supplemental

1. Does the organization maintain documented information that demonstrates that all employees are aware of their impact on product quality and the importance of the activities in achieving, maintaining, and improving quality, including customer requirements and the risks involved for the customer with nonconforming product?

7.3.2 Awareness: employee motivation and empowerment

1. Does the organization maintain a documented process to motivate employees to achieve quality objectives, to make continual improvements, and to create an environment that promotes innovation?
2. Does the process include the promotion of quality and technological awareness throughout the whole organization?

7.5 Documented information
7.5.1 General
7.5.1.1 Documented information: quality management system documentation

1. Is the organization's quality management system documented and does it include a quality manual, which can be a series of documents (electronic or hard copy)?

2. Does the format and structure of the quality manual at the discretion of the organization depend on the organization's size, culture, and complexity?

3. If a series of documents is used, is a list retained of the documents that comprise the quality manual for the organization?

4. Does the quality manual include the scope of the quality management system, including details of and justification for any exclusions?

5. Does the quality manual include documented processes established for the quality management system or reference to them?

6. Does the quality manual include the organization's processes and their sequence and interactions (inputs and outputs), including type and extent of control of any outsourced processes?

7. Does the quality manual include a document (i.e., matrix) indicating where within the organization's quality management system, their customer-specific requirements are addressed?

Note: A matrix of how the requirements of this Automotive QMS Standard are addressed by the organization's processes may be used to assist with linkages of the organization's processes and this Automotive QMS.

7.5.3 Control of documented information

7.5.3.2.1 Record retention

1. Does the organization define, document, and implement a record retention policy?

2. Do the control of records satisfy statutory, regulatory, organizational, and customer requirements?

3. Are production part approvals, tooling records including maintenance and ownership, product and process design records, purchase orders (if applicable), or contracts and amendments retained for the length of time that the product is active for production and service requirements, plus one calendar year, unless otherwise specified by the customer or regulatory agency?

4. Does production part approval documented information include approved product, applicable test equipment records, or approved test data?

7.5.3.2.2 Control of documented information: engineering specifications

1. Does the organization have a documented process describing the review, distribution, and implementation of all customer engineering standards/specifications and related revisions based on customer schedules, as required?

2. Does the organization retain a record of the date on which each change is implemented in production?

3. Does implementation include updated documents?

4. Is review completed within ten working days of receipt of notification of engineering standards/specification changes?

Note: A change in these standards/specifications may require an updated record of customer production part approval when these specifications are referenced on the design record or if they affect documents of the production part approval process, such as control plan and risk analysis (such as FMEAs).

CLAUSE 8: OPERATIONS

8.1 Operational planning and control
    8.1.1 Operational planning and control – supplemental

1. When planning for product realization, are the following topics included: (a) customer product requirements and technical specifications, (b) logistics requirements, (c) manufacturing feasibility, (d) project planning, and (e) acceptance criteria?

8.1.2 Confidentiality

1. Has the organization ensured the confidentiality of customer-contracted products and projects under development, including related product information?

8.2 Requirements for products and services
    8.2.1 Customer communication
    8.2.1.1 Customer communication – supplemental

1. Is written or verbal communication in the language agreed with the customer?
2. Does the organization have the ability to run ransomware analysis and communicate necessary information, including data in a customer-specified computer language and format, e.g., computer-aided design data and electronic data interchange?

8.2.2 Determining the requirements for products and services
    8.2.2.1 Determining the requirements for products and services – supplemental

1. Do these requirements include recycling, environmental impact, and characteristics identified as a result of the organization's knowledge of the product and manufacturing processes?
2. Does compliance to any statutory and regulatory requirement related to product include all applicable government, safety, and environmental regulations related to acquisition, storage, handling, recycling, elimination, or disposal of material?

8.2.3 Review of the requirements for products and services
    8.2.3.1.1 Review of the requirements for products and services – supplemental

1. Does the organization retain documented evidence of a customer-authorized waiver for the requirements stated in ISO 9001, Section 8.2.3.1, for a formal review?

8.2.3.1.2 Customer-designed special characteristics

1. Does the organization conform to customer requirements for designation, approval documentation, and control of special characteristics?

8.2.3.1.3 Requirements for products and services; organizational manufacturing feasibility

1. Does the organization utilize a multidisciplinary approach to conduct an analysis to determine if it is feasible that the organization's manufacturing processes are capable of consistently producing product that meets all of the engineering and capacity requirements specified by the customer?
2. Does the organization conduct this feasibility analysis for any manufacturing or product technology new to the organization and for any changed manufacturing process or product design?
3. Additionally, does the organization validate through production runs, benchmarking studies, or other appropriate methods, their ability to make product to specifications at the required rate?

8.3 Design and development of products and services
   8.3.1 General
   8.3.1.1 Design and development of products and services – supplemental

1. Does the requirement of product and manufacturing process design and development focusses on error prevention rather on detection?
2. Does the organization's documents its design and development processes?

8.3.2 Design and development planning
   8.3.2.1 Design and development planning – supplemental

1. Does the organization ensure that design and development planning include all affected stakeholders within the organization and, as appropriate, its supply chain?
2. While doing the design and development planning, does the organization use as a multidisciplinary approach which includes (a) project management (e.g., APQP or VDA – RGA); (b) product and manufacturing process design activities (e.g., DFM and DFA), such as consideration of the use of alternative designs and manufacturing processes; (c) development and review of product design risk analysis (FMEAs), including actions to reduce potential risks; and (d) development and review of manufacturing process risk analysis (e.g., FMEAs, process flows, control plans, and standard work instructions)?

Note: A multidisciplinary approach typically includes the organization's design, manufacturing, engineering, quality, production, purchasing, supplier, maintenance, and other appropriate functions.

8.3.2.2 Product design skills

1. Does the organization ensure that personnel with product design responsibility are competent to achieve design requirements and are skilled in applicable product design tools and techniques?
2. Are applicable tools and techniques identified by the organization?

8.3.2.3 Development of products with embedded software

1. Does the organization use a process for quality assurance for their products with internally developed embedded software?
2. Is a software development assessment methodology utilized to assess the organization's software development process?
3. Using prioritization based on risk and potential impact to the customer, does the organization retain documented information of a software development capability self-assessment?
4. Does the organization include software development within the scope of their internal audit program?

8.3.3 Design and development inputs
  8.3.3.1 Product design input

1. Does the organization identify, document, and review product design input requirements as a result of contract review?
2. Do product design input requirements include product specifications including but not limited to special characteristics?
3. Do product design input requirements include boundary and interface requirements?
4. Do product design input requirements include identification, traceability, and packaging?
5. Do product design input requirements include consideration of design alternatives?
     Note: One approach for considering design alternatives is the use of trade-off curves.
6. Do product design input requirements include assessment of risks with the input requirements and the organization's ability to mitigate/manage the risks, including from the feasibility analysis?
7. Do product design input requirements include targets for conformity to product requirements including preservation, reliability, durability, serviceability, health, safety, environmental, development timing, and cost?
8. Do product design input requirements include applicable statutory and regulatory requirements of the customer-identified country of destination, if provided?
9. Do product design input requirements include embedded software requirements?

10. Does the organization have a process to deploy information gained from previous design projects, competitive product analysis (benchmarking), supplier feedback, internal input, field data, and other relevant sources for current and future projects of a similar nature?

8.3.3.2 Manufacturing process design input

1. Does the organization identify, document, and review manufacturing process design input requirements?
2. Does the manufacturing process design input requirements including but not limited to the following: (a) product design output data including special characteristics; (b) targets for productivity, process capability, timing, and cost; (c) manufacturing technology alternatives; (d) customer requirements, if any; (e) experience from previous developments; (f) new materials; (g) product handling and ergonomic requirements; and (h) design for manufacturing and design for assembly?
3. Does the manufacturing process design include the use of error-proofing methods to a degree appropriate to the magnitude of the problems and commensurate with the risks encountered?

8.3.3.3 Special characteristics

1. Does the organization use a multidisciplinary approach to establish, document, and implement its process to identify special characteristics, including those determined by the customer and the risk analysis performed by the organization?
2. Does it include documentation of all special characteristics in the drawings (as required), risk analysis (such as FMEA), control plans, and standard work/operator instructions; and special characteristics identified with specific markings and cascaded through each of these documents?
3. Does identification of special characteristics include development of control and monitoring strategies for special characteristics of products and production processes?
4. Does identification of special characteristics include customer-specified approvals, when required?
5. Does identification of special characteristics include compliance with customer-specified definitions and symbols or the organization's equivalent symbols or notations, as defined in a symbol conversion table?
6. Is the symbol conversion table submitted to the customer, if required?

8.3.4 Design and development controls
8.3.4.1 Monitoring

1. Are measurements at specified stages during the D&D of products and processes defined, analyzed, and reported with summary results as an input to management review?

2. When required by the customer, are measurements of the product and process development activity reported to the customer at stages specified, or agreed to, by the customer?
3. When appropriate, do these measurements include quality risks, costs, lead times, critical paths, and other measurements?

8.3.4.2 Design and development validation

1. Is D&D validation performed in accordance with customer requirements, including any applicable industry and governmental agency-issued regulatory standards?
2. Is the timing of D&D validation planned in alignment with customer-specified timing, as applicable?
3. Where contractually agreed with the customer, does this include evaluation of the interaction of the organization's product, including embedded software, within the system of the final customer's product?

8.3.4.3 Prototype program

1. When required by the customer, does the organization have a prototype program and control plan?
2. Does the organization use, whenever possible, the same suppliers, tooling, and manufacturing processes as used in production?
3. Are all performance-testing activities monitored for timely completion and conformity to requirements?
4. When services are outsourced, does the organization include the type and extent of control in the scope of its quality management system to ensure that outsourced services conform to requirements?

8.3.4.4 Product approval process

1. Does the organization establish, implement, and maintain a product and manufacturing approval process conforming to requirements defined by the customer?
2. Does the organization approve externally provided products and services per ISO 9001, Section 8.4.3 (Information for external provider), prior to submission of their part approval to the customer?
3. Does the organization obtain documented product approval prior to shipment, if required by the customer? Are records of such approval retained?
4. Are records of such approval retained?

Note: Product approval should be subsequent to the verification of the manufacturing process
   8.3.5 Design and development outputs
   8.3.5.1 Design and development outputs – supplemental

1. Is the product design output expressed in terms that can be verified and validated against product design input requirements?
2. Does the product design output include design risk analysis (Design FMEA)?
3. Does the product design output include reliability study results?
4. Does the product design output include product special characteristics?
5. Does the product design output include results of product design error-proofing, such as DFSS, DFMA, and FTA?
6. Does the product design output include product definition including 2D drawing, 3D models, technical data packages, product manufacturing information, and geometric dimensioning and tolerancing (GD&T)?
7. Does the product design output include product design review results?
8. Does the product design output include service diagnostic guidelines and repair and serviceability instructions?
9. Does the product design output include service part requirements?
10. Does the product design output include packaging and labeling requirements for shipping?
11. Does the interim design outputs include any engineering problems being resolved through a trade-off process?

8.3.5.2 Manufacturing process design output

1. Does the organization document the manufacturing process design output in a manner that enables verification against the manufacturing process design inputs?
2. Does the organization verify the outputs against manufacturing process design input requirements?
3. Does the manufacturing process design output include specifications and drawings?
4. Does the manufacturing process design output include special characteristics for product and manufacturing process?
5. Does the manufacturing process design output include identification of process input variables that impact characteristics?
6. Does the manufacturing process design output include tooling and equipment for production and control, including capability studies of equipment and process?
7. Does the manufacturing process design output include manufacturing process flowcharts/layout, including linkage of product, process, and tooling?
8. Does the manufacturing process design output include capacity analysis?
9. Does the manufacturing process design output include manufacturing process FMEA?
10. Does the manufacturing process design output include maintenance plans and instructions?
11. Does the manufacturing process design output include control plan?
12. Does the manufacturing process design output include standard work and work instructions?

13. Does the manufacturing process design output include process approval acceptance criteria?
14. Does the manufacturing process design output include data for quality, reliability, maintainability, and measurability?
15. Does the manufacturing process design output include results of error-proofing identification and verification, as appropriate?
16. Does the manufacturing process design output include methods of rapid detection, feedback, and correction of product/manufacturing process nonconformities?

8.3.6 Design and development changes
8.3.6.1 Design and development changes – supplemental

1. Does the organization evaluate all design changes after initial product approval, including those proposed by the organization or its suppliers, for potential impact on fit, form, function, performance, and/or durability?
2. Are these changes validated against customer requirements and approved internally, prior to production implementation?
3. If required by the customer, does the organization obtain documented approval, or a documented waiver, from the customer prior to production implementation?
4. For products with embedded software, does the organization document the revision level of software and hardware as part of the change record?

8.4 Control of extremely provided processes, products, and services
8.4.1 General
8.4.1.1 General – supplemental

1. Does the organization include all products and services that affect customer requirements such as subassembly, sequencing, sorting, rework, and calibration services in the scope of their definition of externally provided products, processes, and services?

8.4.1.2 Supplier selection process

1. Does the organization have a documented supplier selection process?
2. Does the selection process include an assessment of the selected supplier's risk to product conformity and uninterrupted supply of the organization's product to the customers?
3. Does the selection process include relevant quality and delivery performance?
4. Does the selection process include an evaluation of the supplier's quality management system?
5. Does the selection process include multidisciplinary decision-making?
6. Does the selection process include an assessment of software development capabilities, if applicable?
7. Are other supplier selection criteria considered including the following: volume of automotive business (absolute and as a percentage of total business);

financial stability; purchased product, material, or service complexity; required technology (product or process); adequacy of available resources (e.g., people, infrastructure); D&D capabilities (including project management); manufacturing capability; change management process; business continuity planning (e.g., disaster preparedness, contingency planning); logistics process; and customer service

8.4.1.3 Customer-directed sources (also known as *Directed–Buy*)

1. When specified by the customer, does the organization purchase products, materials, or services from customer-directed sources?
2. Are all requirements of Section 8.4 (except the requirements in IATF 16949, Section 8.4.1.2) applicable to the organization's control of customer-directed sources unless specific agreements are otherwise defined by the contract between the organization and the customer?

8.4.2 Type and extent of control
8.4.2.1 Type and extent of control – supplemental

1. Does the organization have a documented process to identify outsourced processes and to select the types and extent of controls used to verify conformity of externally provided products, processes, and services to internal (organizational) and external customer requirements?
2. Does the process include the criteria and actions to escalate or reduce the types and extent of controls and development activities based on supplier performance and assessment of product, material, or service risks?

8.4.2.2 Statutory and regulatory requirements

1. Does the organization document their process to ensure that purchased products, processes, and services conform to the current applicable statutory and regulatory requirements in the country of receipt, the country of shipment, and the customer-identified country of destination, if provided?
2. If the customer defines special controls for certain products with statutory and regulatory requirements, does the organization ensure they are implemented and maintained as defined, including at suppliers?

8.4.2.3 Supplier quality management system development

1. Does the organization require their suppliers of automotive products and services to develop, implement, and improve a quality management system certified to ISO 9001, unless otherwise authorized by the customer, with the ultimate objective of becoming certified to this Automotive QMS Standard?
2. Unless otherwise specified by the customer, is the following sequence applied to achieve this requirement:

- Compliance to ISO 9001 through second-party audits?
- Certification to ISO 9001 through third-party audits; unless otherwise specified by the customer, do suppliers to the organization demonstrate conformity to ISO 9001 by maintaining a third-party certification issued by a certification body bearing the accreditation mark of a recognized IAF MLA (International Accreditation Forum Multilateral Recognition Arrangement) member and where the accreditation body's main scope includes management system certification to ISO/IEC 17021?
- Certification to ISO 9001 with compliance to other customer-defined QMS requirements (such as Minimum Automotive Quality Management System Requirements for Sub-Tier Suppliers [MAQMSR] or equivalent) through second-party audits?
- Certification to ISO 9001 with compliance to IATF 16949 through second-party audits?
- Certification to 16949 through third-party audits (valid third-party certification of the supplier to IATF 16949 by an IATF-recognized certification body)?

8.4.2.3.1 Automotive product-related software or automotive products with embedded software

1. Does the organization require their suppliers of automotive product-related software, or automotive products with embedded software, to implement and maintain a process for software quality assurance for their products?
2. Is a software development assessment methodology utilized to assess the supplier's software development process?
3. Using prioritization based on risk and potential impact to the customer, does the organization require the supplier to retain documented information of a software development capability self-assessment?

8.4.2.4 Supplier monitoring

1. Does the organization have a documented process and criteria to evaluate supplier performance in order to ensure conformity of externally provided products, processes, and services to internal and external customer requirements?
2. At a minimum, are the following supplier performance indicators monitored:
   - Delivered product conformity to requirements?
   - Customer disruptions at the receiving plant, including yard holds and stop ships?
   - Delivery schedule performance?
   - Number of occurrences of premium freight?
3. If provided by the customer, does the organization also include the following, as appropriate, in their supplier performance monitoring:
   - Special status customer notifications related to quality or delivery issues?
   - Dealer returns, warranty, field actions, and recalls?

8.4.2.4.1 Second-party audits

1. Does the organization include a second-party audit process in their supplier management approach? (Second-party audits may be used for the following: (a) supplier risk assessment; (b) supplier monitoring; (c) supplier QMS development; (d) product audits; and (e) process audits).
2. Based on a risk analysis, including product safety/regulatory requirements, performance of the supplier, and QMS certification level, at a minimum, does the organization document the criteria for determining the need, type, frequency, and scope of second-party audits? Does the organization retain records of the second-party audit reports?
3. If the scope of the second-party audit is to assess the supplier's quality management system, is the approach consistent with the automotive process approach?

8.4.2.5 Supplier development

1. Does the organization determine the priority, type, extent, and timing of required supplier development actions for its active suppliers?
2. Do determination inputs include performance issues identified through supplier monitoring?
3. Do determination inputs include second-party audit findings?
4. Do determination inputs include third-party quality management system certification status?
5. Do determination inputs include risk analysis?
6. Does the organization implement actions necessary to resolve open (unsatisfactory) performance issues and pursue opportunities for continual improvement?

8.4.3 Information for external providers
   8.4.3.1 Information for external providers – supplemental
   1. Does the organization pass down all applicable statutory and regulatory requirements and special product and process characteristics to their suppliers and require the suppliers to cascade all applicable requirements down the supply chain to the point of manufacture?

8.5 Production and service provision
   8.5.1 Control of production and service provision
   8.5.1.1 Control plan

1. Does the organization develop control plans at the system, subsystem, component, and / or material level for the relevant manufacturing site and all product supplied, including those for processes producing bulk materials as well as parts?
2. Are family control plans acceptable for bulk material and similar parts using a common manufacturing process?

3. Does the organization have a control plan for pre-launch and production that shows linkage and incorporates information from the design risk analysis (if provided by the customer), process flow diagram, and manufacturing process risk analysis outputs (such as FMEA)?
4. Does the organization, if required by the customer, provide measurement and conformity data collected during execution of either the pre-launch or production control plans?
5. Does control plan includes controls used for the manufacturing process control, including verification of job set-ups?
6. Does control plan include first-off/last-off part validation, as applicable?
7. Does control plan include methods for monitoring of control exercised over special characteristics, defined by both the customer and the organization?
8. Does control plan include the customer-required information, if any?
9. Does control plan include specified reaction plan when nonconforming product is detected, and the process becomes statistically unstable or not statistically capable?
10. Does the organization review control plans and update when it has shipped nonconforming product to the customer?
11. Does the organization review control plans and update when any change occurs affecting product, manufacturing process, measurement, logistics, supply sources, production volume changes, or risk analysis (FMEA)?
12. Does the organization review control plans and update after a customer complaint and implementation of the associated corrective action, when applicable?
13. Does the organization review control plans and update at a set frequency based on a risk analysis?
14. If required by the customer, does the organization obtain customer approval after review or revision of the control plan?

8.5.1.2 Standardized work – operator instructions and visual standards

1. Does the organization ensure that standardized work documents are communicated to and understood by the employees who are responsible for performing the work?
2. Is it legible and presented in the language understood by the personnel responsible to follow them?
3. Is it accessible for use at the designated work area?
4. Do the standardized work documents also include rules for operator safety?

8.5.1.3 Verification of job set-ups

1. Does the organization verify job set-ups when performed, such as an initial run of a job, material changeover, or job change that requires a new set-up?
2. Does the organization maintain documented information for set-up personnel?
3. Does the organization use statistical methods of verification, where applicable?

4. Does the organization perform first-off/last-off part validation, as applicable; where appropriate, are first-off parts retained for comparison with the last-off parts; where appropriate, are last-off parts retained for comparison with first-off parts in subsequent runs?
5. Does the organization retain records of process and product approval following set-up and first-off/last-off part validations?

8.5.1.4 Verification after shutdown

1. Does the organization define and implement the necessary actions to ensure product compliance with requirements after a planned or unplanned production shutdown period?

8.5.1.5 Total productive maintenance

1. Does the organization develop, implement, and maintain a documented total productive maintenance system?
2. Does the system include identification of process equipment necessary to produce conforming product at the required volume?
3. Does the system include availability of replacement parts for the equipment identified?
4. Does the system include provision of resource for machine, equipment, and facility maintenance?
5. Does the system include packaging and preservation of equipment, tooling, and gauging?
6. Does the system include applicable customer-specific requirements?
7. Does the system include documented maintenance objectives, for example, OEE, MTBF (Mean Time between Failure), MTTR (Mean Time to Repair), and Preventive Maintenance Compliance Metrics?
8. Does performance to the maintenance objectives form an input into management review?
9. Does the system include regular review of maintenance plan and objectives and a documented action plan to address corrective actions where objectives are not achieved?
10. Does the system include use of preventive maintenance methods?
11. Does the system include use of predictive maintenance methods, as applicable?
12. Does the system include periodic overhaul?

8.5.1.6 Management of production tooling and manufacturing, test, inspection tooling and equipment

1. Does the organization provide resources for tool and gauge design, fabrication, and verification activities for production and service materials and for bulk materials, as applicable?

2. Does the organization establish and implement a system for production tooling management, whether owned by the organization or the customer?
3. Does the production tooling management include maintenance and repair facilities and personnel?
4. Does the production tooling management include storage and recovery?
5. Does the production tooling management include set-up and tool-change programs for perishable tools?
6. Does the production tooling management include tool design modification documentation, including engineering change level of the product?
7. Does the production tooling management include tool modification and revision to documentation?
8. Does the production tooling management include tool identification, such as serial or asset number; the status, such as production, repair, or disposal; ownership; and location?
9. Does the organization verify that customer-owned tools, manufacturing equipment, and test/inspection equipment are permanently marked in a visible location so that the ownership and application of each item can be determined?
10. Does the organization implement a system to monitor these activities if any work is outsourced?

8.5.1.7 Production scheduling

1. Does the organization ensure that production is scheduled in order to meet customer orders/demands such as Just-In-Time (JIT) and is supported by an information system that permits access to production information at key stages of the process and is order driven?
2. Does the organization include relevant planning information during production scheduling, e.g., customer orders, supplier on-time delivery performance, capacity, shared loading (multi-part station), lead time, inventory level, preventive maintenance, and calibration?

8.5.2 Identification and traceability
8.5.2.1 Identification and traceability – supplemental

1. Does the organization implement identification and traceability processes to support identification of clear start and stop points for product received by the customer or in the field that may contain quality- and/or safety-related nonconformities?
2. Does the organization conduct an analysis of internal, customer, and regulatory traceability requirements for all automotive products, including developing and documenting traceability plans, based on the levels of risk or failure severity for employees, customers, and consumers? By the way, for those of you who are looking for 18 wheeler truck accident lawyers, visit sloanfirm. com. They can assist you in any legal help especially in car accidents.

3. Do these plans define the appropriate traceability systems, processes, and methods by product, process, and manufacturing location?
4. Do these plans enable the organization to identify nonconforming and/or suspect product?
5. Do these plans enable the organization to segregate nonconforming and/or suspect product?
6. Do these plans ensure the ability to meet the customer and/or regulatory response time requirements?
7. Do these plans ensure documented information is retained in the format (electronic, hardcopy, archive) that enables the organization to meet the response time requirements?
8. Do these plans ensure serialized identification of individual products, if specified by the customer or regulatory standards?
9. Do these plans ensure the identification and traceability requirements are extended to externally provided products with safety/regulatory characteristics?

8.5.4 Preservation
8.5.4.1 Preservation – supplemental

1. Does preservation include identification, handling, contamination control, packaging, storage, transmission or transportation, and protection?
2. Does preservation apply to materials and components from external and/or internal providers from receipt through processing, including shipment and until delivery to/acceptance by the customer?
3. In order to detect deterioration, does the organization assess at appropriate planned intervals the condition of product in stock, the place/type of storage container, and the storage environment?
4. Does the organization use an inventory management system to optimize inventory turns over time and ensure stock rotation, such as "first in first out" (FIFO)?
5. Does the organization ensure that obsolete product is controlled in a manner similar to that of nonconforming product?
6. Do organizations comply with preservation, packaging, shipping, and labeling requirements as provided by their customers?

8.5.5 Post-delivery activities
8.5.5.1 Feedback of information from service

1. Does the organization ensure that a process for communication of information on service concerns to manufacturing, material handling, logistics, engineering, and design activities is established, implemented, and maintained?
2. Is the organization aware of nonconforming products and materials that may be identified at the customer location or in the field?
3. Where applicable does "service concerns" include the results of field failure test analysis?

8.5.5.2 Service agreement with customer

1. When there is a service agreement with the customer, does the organization verify that the relevant service centers comply with applicable requirements?
2. Does the organization verify the effectiveness of any special purpose tools or measurement equipment?
3. Does the organization ensure that all service personnel are trained in applicable requirements?

8.5.6 Control of change
   8.5.6.1 Control of change – supplemental

1. Does the organization have a documented process to control and react to changes that impact product realization?
2. Are the effects of any change, including those changes caused by the organization, the customer, or any supplier, assessed?
3. Does the organization define verification and validation activities to ensure compliance with customer requirements?
4. Does the organization validate changes before implementation?
5. Does the organization document the evidence of related risk analysis?
6. Does the organization retain records of verification and validation?
7. Do changes, including those made at suppliers, require a production trial run for verification of changes such as changes to part design, manufacturing location, or manufacturing process to validate the impact of any changes on the manufacturing process?
8. When required by the customer, does the organization notify the customer of any planned product realization changes after the most recent product approval?
9. When required by the customer, does the organization obtain documented approval, prior to implementation of the change?
10. When required by the customer, does the organization complete additional verification or identification requirements, such as production trial run and new product validation?

8.5.6.1.1 Temporary change of process control

1. Does the organization identify, document, and maintain a list of the process controls, including inspection, measuring, test, and error-proofing devices, that includes the primary process control and the approved back-up or alternate methods?
2. Does the organization document the process that manages the use of alternate control methods?
3. Does the organization include in this process, based on risk analysis (such as FMEA), severity, and the internal approvals to be obtained prior to production implementation of the alternate control method?

4. Before shipping product that was inspected or tested using the alternate method, if required, does the organization obtain approval from the customer(s)?
5. Does the organization maintain and periodically review a list of approved alternate process control methods that are referenced in the control plan?
6. Are standard work instructions available for each alternate process control method?
7. Does the organization review the operation of alternate process controls on a daily basis, at a minimum, to verify implementation of standard work with the goal to return to the standard process as defined by the control plan as soon as possible? Example methods include but are not limited to the following:
   - Quality focused audits (e.g., LPAs, as applicable)
   - Daily leadership meetings.
8. Is restart verification documented for a defined period based on severity and confirmation that all features of the error-proofing device or process are effectively reinstated?
9. Does the organization implement traceability of all product produced while any alternate process control devices or processes are being used (e.g., verification and retention of first piece and last piece from every shift)?

8.6 Release of products and services
   8.6.1 Release of products and services – supplemental

1. Does the organization ensure that the planned arrangements to verify that the product and service requirements have been met encompass the control plan and are documented as specified in the control plan?
2. Does the organization ensure that the planned arrangements for the initial release of products and services encompass product or service approval?
3. Does the organization ensure that product or service approval is accomplished after changes following initial release, according to ISO 9001, Section 8.5.6?

8.6.2 Layout inspection and functional testing

1. Is a layout inspection and a functional verification to applicable customer engineering material and performance standards performed for each product as specified in the control plans?
2. Are results available for customer review?

Note 1: Layout inspection is the complete measurement of all product dimensions shown on the design record(s).
Note 2: The frequency of layout inspection is determined by the customer.
   8.6.3 Appearance items

1. For organizations manufacturing parts designated by the customer as "appearance items," does the organization provide appropriate resources, including lighting, for evaluation?

2. Does the organization provide masters for color, grain, gloss, metallic brilliance, texture, distinctness of image (DOI), and haptic technology, as appropriate?
3. Does the organization provide maintenance and control of appearance masters and evaluation equipment?
4. Does the organization provide verification that personnel making appearance evaluations are competent and qualified to do so?

8.6.4 Verification and acceptance of conformity of externally provided products and services

1. Does the organization have a process to ensure the quality of externally provided processes, products, and services utilizing one or more of the following methods?

Receipt and evaluation of statistical data provided by the supplier to the organization; receiving inspection and/or testing, such as sampling based on performance; second-party or third-party assessments or audits of supplier sites when coupled with records of acceptable delivered product conformance to requirements; part evaluation by a designated laboratory; and another method agreed with the customer?

8.6.5 Statutory and regulatory conformity

1. Prior to release of externally provided products into its production flow, does the organization confirm and is it able to provide evidence that externally provided processes, products, and services conform to the latest applicable statutory, regulatory, and other requirements in the countries where they are manufactured and in the customer-identified countries of destination, if provided?

8.6.6 Acceptance criteria

1. Is acceptance criteria defined by the organization and, where appropriate or required, approved by the customer?
2. For attributed data sampling, is the acceptance level zero defects?

8.7 Control of nonconforming outputs
8.7.1.1 Customer authorization for concession

1. Does the organization obtain a customer concession or deviation permit prior to further processing whenever the product or manufacturing process is different from that which is currently approved?
2. Does the organization obtain customer authorization prior to further processing for "use as is" and rework dispositions of nonconforming product?
3. If subcomponents are reused in the manufacturing process, is that subcomponent reuse clearly communicated to the customer in the concession or deviation permit?

4. Does the organization maintain a record of the expiration date or quantity authorized under concession?
5. Does the organization also ensure compliance with the original or superseding specifications and requirements when the authorization expires?
6. Is material shipped under concession properly identified on each shipping container (this applies equally to purchased product)?
7. Does the organization approve any requests from suppliers before submission to the customer?

8.7.1.2 Control of nonconforming product – customer-specified process

1. Does the organization comply with applicable customer-specified controls for nonconforming product(s)?

8.7.1.3 Control of suspect product

1. Does the organization ensure that product with unidentified or suspect status is classified and controlled as nonconforming product?
2. Does the organization ensure that all appropriate manufacturing personnel receive training for containment of suspect and nonconforming product?

8.7.1.4 Control of Reworked Product

1. Does the organization utilize risk analysis (such as FMEA) methodology to assess risks in the rework process prior to a decision to rework the product?
2. If required by the customer, does the organization obtain approval from the customer prior to commencing rework of the product?
3. Does the organization have a documented process for rework confirmation in accordance with the control plan or other relevant documented information to verify compliance to original specifications?
4. Are instructions for disassembly or rework, including re-inspection and traceability requirements, accessible to and utilized by the appropriate personnel?
5. Does the organization retain documented information on the disposition of reworked product including quantity, disposition, disposition date, and applicable traceability information?

8.7.1.5 Control of repaired product

1. Does the organization utilize risk analysis (such as FMEA) methodology to assess risks in the repair process prior to a decision to repair the product?
2. Does the organization obtain approval from the customer before commencing repair of the product?
3. Does the organization have a documented process for repair confirmation in accordance with the control plan or other relevant documented information?

4. Are instructions for disassembly or repair, including re-inspection and traceability requirements, accessible to and utilized by the appropriate personnel?
5. Does the organization obtain a documented customer authorization for concession for the product to be repaired?
6. Does the organization retain documented information on the disposition of repaired product including quantity, disposition, disposition date, and applicable traceability information?

8.7.1.6 Customer notification

1. Does the organization immediately notify the customers in the event that nonconforming product has been shipped?
2. Is initial communication followed with detailed documentation of the event?

8.7.1.7 Nonconforming product disposition

1. Does the organization have a documented process for disposition of nonconforming product not subject to rework or repair?
2. For product not meeting requirements, does the organization verify that the product to be scrapped is rendered unusable prior to disposal?
3. The organization shall not divert nonconforming product to service or other use without prior customer approval.

CLAUSE 9: PERFORMANCE EVALUATION

9.1 Monitoring, measurement, analysis, and evaluation
  9.1.1 General
  9.1.1.1 Monitoring and measurement of manufacturing processes

1. Does the organization perform process studies on all new manufacturing (including assembly or sequencing) processes to verify process capability and to provide additional input for process control, including those for special characteristics?
2. For manufacturing processes where it may not be possible to demonstrate product compliance through process capability, are alternate methods such as batch conformance to specification used?
3. Does the organization maintain manufacturing process capability or performance results as specified by the customer's part approval process requirements?
4. Does the organization verify that the process flow diagram, PFMEA, and control plan are implemented?
5. Does the organization adhere to the following? (a) measurement techniques, (b) sampling plans, (c) acceptance criteria, (d) records of actual measurement values and/or test results for variable data, and (e) reaction plans and escalation process when acceptance criteria are not met

6. Are significant process events, such as tool change or machine repair, recorded and retained as documented information?
7. Does the organization initiate a reaction plan indicated on the control plan and evaluated for impact on compliance to specifications for characteristics that are either not statistically capable or are unstable?
8. Do these reaction plans include containment of product and 100% inspection, as appropriate?
9. Is a corrective action plan developed and implemented by the organization indicating specific actions, timing, and assigned responsibilities to ensure that the process becomes stable and statistically capable?
10. Does the organization review the plans with and approved by the customer, when required?
11. Does the organization maintain records of effective dates of process changes?

9.1.1.2 Identification of statistical tools

1. Does the organization determine the appropriate use of statistical tools?
2. Does the organization verify that appropriate statistical tools are included as part of the advanced product quality planning (or equivalent) process and included in the design risk analysis (such as DFMEA) (where applicable), the process risk analysis (such as PFMEA), and the control plan?

9.1.1.3 Application of statistical concepts

1. Are statistical concepts, such as variation, control (stability), process capability, and the consequences of over-adjustment, understood and used by employees involved in the collection, analysis, and management of statistical data?

9.1.2 Customer satisfaction
9.1.2.1 Customer satisfaction – supplemental

1. Is customer satisfaction with the organization monitored through continual evaluation of internal and external performance indicators to ensure compliance to the product and process specifications and other customer requirements?
2. Are performance indicators based on objective evidence and include but not limited to the following: (a) delivered part quality performance?
3. Do performance indicators include customer disruptions?
4. Do performance indicators include field returns, recalls, and warranty (where applicable)?
5. Do performance indicators include delivery schedule performance (including incidents of premium freight)?
6. Do performance indicators include customer notifications related to quality or delivery issues, including special status?

7. Does the organization monitor the performance of manufacturing processes to demonstrate compliance with customer requirements for product quality and process efficiency?
8. Does the organization monitor the performance of manufacturing processes to demonstrate compliance with customer requirements for product quality and process efficiency?
9. Does the organization monitor the performance of manufacturing processes to demonstrate compliance with customer requirements for product quality and process efficiency?
10. Does the organization record analytical results, and retain and control these records?

9.1.3 Analysis and evaluation
 9.1.3.1 Prioritization

1. Are trends in quality and operational performance compared with progress towards objectives and lead to action to support prioritization of actions for improving customer satisfaction?

9.2 Internal audit
 9.2.2.1 Internal audit program

1. Does the organization have a documented internal audit process?
2. Does the process include the development and implementation of an internal audit program that covers the entire quality management system including quality management system audits, manufacturing process audits, and product audits?
3. Is the audit program prioritized based upon risk, internal and external performance trends, and criticality of the processes?
4. Where the organization is responsible for software development, does the organization include software development capability assessments in their internal audit program?
5. Is the frequency of audits reviewed and, where appropriate, adjusted based on occurrence of process changes, internal and external nonconformities, and/or customer complaints?
6. Is the effectiveness of the audit program reviewed as a part of management review?

9.2.2.2 Quality management system audit

1. Does the organization audit all quality management system processes over each 3-year calendar period, according to an annual program, using the process approach to verify compliance with this Automotive QMS Standard?
2. Integrated with these audits, does the organization sample customer-specific quality management system requirements for effective implementation?

9.2.2.3 Manufacturing process audit

1. Does the organization audit all manufacturing processes over each 3-year calendar period to determine their effectiveness and efficiency using customer-specified required approaches for process audits?
2. Where not defined by the customer, does the organization determine the approach to be used?
3. Within each individual audit plan, is each manufacturer process audited on all shifts where it occurs, including the appropriate sampling of the shift handover?
4. Does the manufacturing process audit include an audit of the effective implementation of the process risk analysis (such as PFMEA), control plan, and associated documents?

9.2.2.4 Product audit

1. Does the organization audit products using customer-specific required approaches at appropriate stages of production and delivery to verify conformity to specified requirements?
2. Where not defined by the customer, does the organization define the approach to be used?

9.3 Management review
    9.3.1 General
    9.3.1.1 Management review – supplemental

1. Is management review conducted at least annually?
2. Is the frequency of management review(s) increased based on risk to compliance with customer requirements resulting from internal or external changes impacting the quality management system and performance-related issues?

9.3.2 Management review inputs
    9.3.2.1 Management review inputs – supplemental

1. Does input to management review include cost of poor quality (cost of internal and external nonconformance)?
2. Does input to management review include measures of process effectiveness?
3. Does input to management review include measures of process efficiency?
4. Does input to management review include product conformance?
5. Does input to management review include assessments of manufacturing feasibility made for changes to existing operations and for new facilities or new product?
6. Does input to management review include customer satisfaction?
7. Does input to management review include review of performance against maintenance objectives?

8. Does input to management review include warranty performance where applicable?
9. Does input to management review include review of customer scorecards where applicable?
10. Does input to management review include identification of potential field failures identified through risk analysis (such as FMEA)?
11. Does input to management review include actual field failures and their impact on safety or the environment?

9.3.3 Management review outputs
9.3.3.1 Management review outputs – supplemental

1. Does top management document and implement an action plan when customer performance targets are not met?

CLAUSE 10: IMPROVEMENT

10.2 Nonconformity and corrective action
10.2.3 Problem-solving

1. Does the organization have a documented process for problem-solving?
2. Has the organization defined approaches for various types and scale of problems (*e.g.*, new product development, current manufacturing issues, field failures, audit findings)?
3. Does the process include containment, interim actions, and related activities necessary for control of nonconforming outputs?
4. Does it include root cause analysis, methodology used, analysis, and results?
5. Does it include implementation of systemic corrective actions, including consideration of the impact on similar processes and products?
6. Does the organization have a system to verify the effectiveness of implemented corrective actions?
7. Does the organization review and, where necessary, update the appropriate documented information (*e.g.*, PFMEA, control plan)?
8. Where the customer has specified prescribed processes, tools, or systems for problem-solving, does the organization use those processes, tools, or systems, unless otherwise approved by the customer?

10.2.4 Error-proofing

1. Does the organization have a documented process to determine the use of appropriate error-proofing methodologies?
2. Are details of the method used documented in the process risk analysis (such as PFMEA) and are test frequencies documented in the control plan?
3. Does the process include the testing of error-proofing devices for failure or simulated failure?
4. Are records maintained?

5. Are challenge parts, when used, identified, controlled, verified, and calibrated where feasible?
6. Do error-proofing device failures have a reaction plan?

10.2.5 Warranty management systems

1. When the organization is required to provide warranty for their products, does the organization implement a warranty management process?
2. Does the organization include in the process a method for warranty part analysis, including NTF (no trouble found)?
3. When specified by the customer, does the organization implement the required warranty management process?

10.2.6 Customer complaints and field failure test analysis

1. Does the organization perform analysis on customer complaints and field failures, including any returned parts, and does it initiate problem-solving and corrective action to prevent recurrence?
2. Where requested by the customer, does this include analysis of the interaction of embedded software of the organization's product within the system of the final customer's product?
3. Does the organization communicate the results of testing/analysis to the customer and also within the organization?

10.3 Continual improvement
10.3.1 Continual improvement – supplemental

1. Does the organization have a documented process for continual improvement?
2. Does it include identification of the methodology used, objectives, measurement, effectiveness, and documented information?
3. Does it include a manufacturing process improvement action plan with emphasis on the reduction of process variation and waste?
4. Does it include risk analysis (such as FMEA)?

Note: Continual improvement is implemented once manufacturing processes are statistically capable and stable or when product characteristics are predictable and meet customer requirements.

VDA 6.3

OVERVIEW

The VDA audit approach is quite different than any other audit – and there are many types. An overview relationship of the three most common different audits is shown in Table 5.3.

TABLE 5.3

Relationship between System, Process, and Product Audit

System Audit	Process Audit	Product Audit
Quality system	Product development process and/or serial production	Products or services
Assessment of the completeness and effectiveness of the basic requirements.	Service development process and/or Providing the service	Assessment of quality characteristics
	Assessment of the quality capability for specific products/product groups and their processes	

Process audits may be initiated, *e.g.,* for the following reasons: decreasing process quality, customer complaints, and changes in the production sequence – process insecurities, cost reductions, and internal request. The VDA provides a methodology that addresses *all* these issues, and it does it in a very prescriptive way. It may be used internally and externally across the full quality cycle in the following areas: marketing, development, purchasing (product and/or service), production/service provision, sales/commissioning, customer service/services, recycling, and so on.

So, since the VDA 6.3 is very prescriptive and very specific in the questions that an auditor may pursue, rather than providing a structured checklist, the following selected categories have been chosen to show the reader the depth of the VDA 6.3 standard. For more information, see: https://nimonikapp.com/public_templates/12996-vda-6-3-process-audit-and-checklist-for-the-car-industry. Retrieved on July 28, 2020. The numbers in parentheses indicate the topics on the pages of the standard.

1. Product Development (Design): Product Development Planning (p. 66)

   a. Are the customer requirements available?
   b. Is a product development plan available and are the targets maintained?
   c. Are the resources for the realization of the product development planned?
   d. Have the product requirements been determined and considered?
   e. Has the feasibility been determined based on the available requirements?
   f. Are the necessary personnel and technical conditions for the project process planned/available?

2. Product Development (Design): Realizing Product Development (pp. 66, 75)

   a. Is the design FMEA raised and are improvement measures established?
   b. Is the design FMEA updated in the project process and are the established measures realized?
   c. Is a quality plan prepared?
   d. Are the required releases/qualification records available at the respective times?
   e. Are the required resources available?

3. Process Development: Process Development Planning (pp. 66, 122)

    a. Are the product requirements available?
    b. Is a process development plan available and are the targets maintained?
    c. Are the resources for the realization of serial production planned?
    d. Have the process requirements been determined and considered?
    e. Are the necessary personnel and technical preconditions for the project process planned/available?
    f. Is the process FMEA raised and are improvement measures established?

4. Process Development: Realizing Process Development (pp. 67, 121)

    a. Is the process FMEA updated when amendments are made during the project process and are the established measures implemented?
    b. Is a quality plan prepared?
    c. Are the required releases/qualification records available at the respective times?
    d. Is a pre-production carried out under serial conditions for the serial release?
    e. Are the production and inspection documents available and complete?
    f. Are the required resources available?

5. Suppliers/Input Material (pp. 120, 121)

    a. Are only approved quality capable suppliers used?
    b. Is the agreed quality of the purchased parts guaranteed?
    c. Is the quality performance evaluated and are corrective actions introduced when there are deviations from the requirements?
    d. Are target agreements for continual improvement of products and process made and implemented with the suppliers?
    e. Are the required releases for the delivered serial products available and the required improvement measures implemented?
    f. Are the procedures agreed with the costumer, regarding customer-supplied products, maintained?
    g. Are the stock levels of input material matched to production needs?
    h. Are input materials/internal residues delivered and stored according to their purpose?
    i. Are the personnel qualified for the respective tasks?

6. Production: Personnel/Qualification (pp. 68, 121)

    a. Are the employees given responsibility and authority for monitoring the product/process quality?
    b. Are the employees given responsibility and authority for production equipment and environment?
    c. Are the employees suitable to perform the required tasks and is their qualification maintained?
    d. Is there a personnel plan with a replacement ruling?
    e. Are instruments to increase employee motivation effectively implemented?

7. Production: Material/Equipment (pp. 68, 122)

a. Are the product-specific quality requirements fulfilled with the production equipment/tools?
b. Can the quality requirements be monitored effectively during serial production with the implemented inspection, measuring, and test equipment?
c. Are the work and inspection stations appropriate to the needs?
d. Are the relevant details in the production and inspection documents complete and maintained?
e. Are the necessary auxiliary means available for adjustments?
f. Is an approval for production starts issued and are adjustment details, as well as deviations, recorded?
g. Are the required corrective actions carried out on schedule and checked for effectiveness?

8. Production: Transport/Parts Handling/Storage/Packaging (pp. 66–67, 120)

a. Are the quantities/production lot sizes matched to the requirements and are they purposefully forwarded to the next work station?
b. Are products/components appropriately stored and are the transport means/packaging equipment tuned to the special properties of the product/ components?
c. Are rejects, rework, and adjustment parts, as well as internal residues strictly separated and identified?
d. Is the material and parts flow secured against mix ups/exchanges by mistake and traceability guaranteed?
e. Are tools, equipment and inspection, measuring, and test equipment stored correctly?

9. Production: Fault Analysis/Correction/Continual Improvement (pp. 69, 122)

a. Are quality and process data recorded complete and ready to be evaluated?
b. Are the quality and process data statistically analyzed and are improvement programs derived from this?
c. Are the causes of product and process nonconformities analyzed and the corrective actions checked for the effectiveness?
d. Are processes and products regularly audited?
e. Are product and process subject to continual improvement?
f. Are target parameters available for product and process and is their compliance monitored?

10. Customer Service, Customer Satisfaction, Service (pp. 70, 123)

a. Are customer requirements fulfilled at delivery?
b. Is customer service guaranteed?
c. Are complaints quickly reacted to and the supply of parts secured?

d. Are fault analyses carried out when there are deviations from the quality requirements and are improvement measures implemented?
e. Are the personnel qualified for each task?

The VDA also provides an example for a self-assessment template and can be found on pages 178–181. Here, we provide the reader with some very important items in the process of evaluating that assessment, which can be for internal and/or external auditors.

INDIVIDUAL EVALUATION OF THE QUESTIONS AND PROCESS ELEMENTS

Each question is evaluated with regard to the respective requirements and their consistent achievement in the product development process (service process) and the serial production (service). The evaluation can result in 0, 4, 8, 10 points for each question, whereby the proven compliance with the requirements is the measure for awarding points. For a grading under 10 points corrective actions with deadlines have to be determined. For the detail evaluation of compliance with individual requirements, see pp. 55–58 and 165–168 of the standard. However, in summary the numerical values are

10: Full compliance with requirements.
8: Predominant compliance with requirements; minor nonconformities. Predominant means that more than ¾ of all requirements have proven to be effective and no special risk is given.
6: Partial compliance with requirements, more severe nonconformities.
4: Unsatisfactory compliance with requirements, major nonconformities.
0: No compliance with requirements

The degree of conformity (*Overall compliance – EE*) of a process element is calculated from

$$EE[\%] = \frac{\text{Sum of all points awarded for the respective questions}}{\text{Sum of all possible points of the respective questions}} \times 100\%$$

OVERALL GRADE (PP. 53–54, 60)

VDA has its own grading. In fact, it has two. The first one is a color scheme:

- Red (R): It means that the supplier is not able to receive the VDA recognition.
- Yellow (Y): The approval is conditional.
- Green (G): Fully approved.

The second one is based on a numerical value – 100 points being the maximum. The breakdown of the numerical values is

- A (90–100 points) in full compliance
- B (80 < 90 points) in partial compliance
- C (<80 points) not quality capable.

Downgrading Rules (p. 61)

VDA has the possibility to downgrade the compliance grade based on the following:

- Downgrade from A to B despite an overall score of >90%
- Process elements P5 to P7, achievement level <80%
- At least one question assessed as 4 points
- At least one question assessed as 0 points.

Downgrade from A or B to C despite an overall score of >80%

- Process elements P5 to P7, achievement level <70%
- At least one question assessed as 0 points.

References

AIAG (2005). *Statistical Process Control, SPC*. 2nd ed. Southfield, MI: Automotive Industry Action Group.

AIAG (2008a). *Failure Mode and Effect Analysis, FMEA, FMEA*. 4th ed. Southfield, MI: Automotive Industry Action Group.

AIAG (2008b). *Advanced Product Quality Planning, APQP*. 2nd ed. Southfield, MI: Automotive Industry Action Group.

AIAG (2009). *Production Part Approval Process, PPAP*. 4th ed. Southfield, MI: Automotive Industry Action Group.

AIAG (2010). *Measurement Systems Analysis, MSA*. 4th ed. Southfield, MI: Automotive Industry Action Group.

AIAG/VDA (2019). *Failure Mode and Effect Analysis – FMEA: Design FMEA and Process FMEA Handbook*. 1st ed. Southfield, MI: Automotive Industry Action Group/Verband der Automobilindustrie.

American Society for Quality Control (1993). *Certification Program for Auditors of Quality Systems*. Milwaukee, WI: ASQC.

Anderson, B. and T. Fagerhaug (2000). *Root Cause Analysis*. Milwaukee, WI: Quality Press.

ANSI/ISO/ASQC A8402 (1994). *Quality Vocabulary*. Milwaukee, WI: ASQC.

Brumm, E. (1995). *Managing Records for ISO 9000 Compliance*. Milwaukee, WI: Quality Press.

Chrysler Ford GM (1994). *Quality System Requirements QS 9000*. Southfield MI: AIAG.

Clements, R., S. Sidor, and R. Winters (1995). *Preparing Your Company for QS 9000: A Guide for the Automotive Industry*. Milwaukee, WI: Quality Press.

Continuous Improvement Newsletter (February, 1993). How to prepare your documentation for ISO 9000 registration. *Continuous Improvement Newsletter*. National ISO 9000 Support Group, Caledonia, MI, pp. 1–3.

Ford Motor Company (2011). *FMEA Handbook (with Robustness Linkages). Version 4.2.* Dearborn, MI: Ford Motor Company.

Grant (June 1, 2018). https://www.world-cert.co.uk/iso-450012018/. Retrieved on November 1, 2019.

Grossman (January – February, 1995). ISO 9000 readiness survey. Quality in Manufacturing. https://www.certificationeurope.com/certification/iso-45001-occupational-health-safety/. Retrieved on November 1, 2019.

Gruska, G. and C. Kymal (June 24, 2019). Introducing the AIAG-VDA DFMEA: Understanding the changes. Quality Digest.

IATF 16949: 2016 (revised 2019). *Automotive Quality Management System Standard.* Southfield, MI: AIAG.

Irwin Professional Publishing. (September, 1993). ISO 9000 Survey. *Quality Systems Update.* Burr Ridge, IL: Irwin Professional Publishing. pp. 1–8.

Isidore, C. (February 4, 2015). https://money.cnn.com/2015/02/04/news/companies/gm-earnings-recall-costs/. Retrieved on May 6, 2020.

ISO 10013 (1995). *Guidelines for Developing Quality Manuals*. Milwaukee WI: ASQC.

ISO 14001 (2015). *Environmental Management System*. Zurich, Switzerland: International Organization for Standardization.

ISO 1431 (2013). *Environmental Management: Environmental Performance Evaluation: Guidelines*. Zurich, Switzerland: International Organization for Standardization.

ISO 18001 (2007). *Health and Safety Management Standard.* Zurich, Switzerland: International Organization for Standardization.

ISO 27001 (2013). *Information Security Standard.* Zurich, Switzerland: International Organization for Standardization.

ISO 45001 (2018). *Occupational Health and Safety.* Zurich, Switzerland: International Organization for Standardization.

ISO 9001 (2015). *Quality Systems: Model for Quality Assurance in Design, Development, Production Installation and Servicing.* Milwaukee, WI: ASQC.

ISO 9001: 2015 (November 2018). *Quality Management System.* Geneva, Switzerland: International Organization for Standardization.

ISO 9001 (2018). *Quality Management System.* Zurich, Switzerland: International Organization for Standardization.

ISO 9004 (2018). *Quality Management and Quality System Elements: Guidelines.* Milwaukee, WI: ASQC.

Noe, J. (March 1, 2017). https://www.insightls.com/process-audit-questions-going-from-good-to-excellent/. Retrieved on March 25, 2020.

Scott, G. (June 1, 2018). Standards spotlight: ISO 45001 vs. OSHAS 18001.

Stamatis, D. (1996). *Documenting and Auditing for ISO 9000 and QS 9000: Tools for Ensuring Certification or Registration.* Chicago, IL: Irwin Professional Publishers.

VDA 6.3. (2016). *Quality Management in the Automotive Industry: Process Audit.* Part 3.3rd rev. ed. Berlin, Germany: Verband der Automobilindustrie.

Wilson, L. (1996). *Eight Step Process to Successful ISO 9000 Implementation: A Quality Management System Approach.* Milwaukee, WI: Quality Press.

Printed in the United States
by Baker & Taylor Publisher Services